A Working Guide
to Process Equipment

Chemical Engineering Books

A Working Guide to Process Equipment

How Process Equipment Works

Norman P. Lieberman
Elizabeth T. Lieberman

Process Chemicals, Inc.
Metairie, Louisiana

Boston, Massachusetts   Burr Ridge, Illinois
Dubuque, Iowa   Madison, Wisconsin   New York, New York
San Francisco, California   St. Louis, Missouri

Library of Congress Cataloging-in-Publication Data

Lieberman, Norman P.
    A working guide to process equipment / Norman P. Lieberman.
Elizabeth T. Lieberman.
       p.   cm.
    Includes index.
    ISBN 0-07-038075-9
    1. Chemical plants—Equipment and supplies. I. Lieberman,
Elizabeth T. II. Title.
TP157.L475   1997
660′.283—dc20
                                    96-30959
                                        CIP

McGraw-Hill

A Division of The McGraw·Hill Companies

6 7 8 9 BKM BKM 9 0 9 8 7 6 5 4 3 2 1

ISBN 0-07-038075-9

*The sponsoring editor for this book was Zoe G. Foundotos, the editing
supervisor was Paul R. Sobel, and the production supervisor was
Suzanne W. B. Rapcavage. It was set in Century Schoolbook by
Renee Lipton of McGraw-Hill's Professional Book Group
composition unit.*

This book is dedicated to our parents:

Elizabeth and Tom Holmes, innovative engineers, courageous under fire at war and in peace.

Mary and Lou Lieberman whose enduring strength and fortitude have been little noted, but long remembered.

Contents

Letter to Readers: How to Read This Book

November 1, 1996

Subject: How Process Equipment Works

Dear Reader:

Thank you for buying our book. We worked very hard writing it, and we appreciate your vote of confidence.

No normal person is going to read this book for fun or relaxation. It is a work book, for working people. You purchased it with the hope and faith that it can help you do a better job. You opened it with the expectation that you can read it with comprehension.

Well, we won't let you down. But, let's make a deal. We promise you that even though this is a technical book, you can read it easily, without pain, but with comprehension. After you read it, you will definitely be a better process operator or engineer. Your part of the deal is to read the whole book. This is not a reference or source book. All the chapters are tied together by threads of logic. You will really find it easier to grasp this logic, if you read the chapters in sequence.

A few of the words in the text are italicized. These words are explained in the Glossary, at the back of the book.

Please feel free to give us a call if there is some point you would like to discuss, or a process question you wish to ask.

Sincerely,

Norm Lieberman
Chemical Engineer

Liz Lieberman
Chemical Engineer

Suite 267
5000A W. Esplanade
Metairie, LA 70006, U.S.A.
(504) 887-7714
Fax (504) 456-1835

Introduction

In 1983, I started teaching a three day process equipment troubleshooting seminar, to chemical engineers and experienced plant operators in the petroleum refining and chemical process industry. Since the inception of the seminar, in excess of 5000 men and women have attended the classes. The seminar is largely based on my 30 years experience in field troubleshooting, and process unit revamp design.

I have taught hundreds of seminars explaining how pumps, compressors, heat exchangers, distillation towers, steam jets, fired heaters, and steam turbines mal-function. I have explained to thousands of chemical engineers how to design trays, and modify tube bundles for improved performance. More thousands of operators have listened to me expound as to how and why cavitation damages pump mechanical seals. And throughout these lectures, one common thread has emerged.

The general knowledge as to how process equipment really functions, is disappearing from the process industries. This is not only my opinion, but the general view of senior technical managers, in many large corporations.

Chemical process equipment is basically the same now, as it was in the 1930's, or at least the 1950's. The trays, K.O. drums, compressors, heaters, steam systems have not - and probably will not change. The fundamental nature of process equipment operation has been well established for a very long time. Modern methods of computer control, and process design have not, and cannot, change the basic performance of the bulk of process equipment. These tools just seem to have made learning about the working of the equipment more difficult.

The chemical engineer has traditionally been the guardian of process knowledge. So, one would suppose that if fundamental process knowledge is vanishing, the origin of the problem may lie in our universities. Perhaps, there is less of that, "hands-on approach," to problems, with the advent of the "P.C." or perhaps there are just fewer people around to teach us. No one really knows.

But in this book, we have gone back to the very simplest basis for understanding process equipment. In every chapter we have said, "Here is how the equipment behaves in the field, and this is why." We have shown how to do simple technical calculations. The guiding idea of our book is that it is better to have a working knowledge of a few simple ideas, than a superficial knowledge of many complex theoretical subjects.

The original three day troubleshooting seminar, has now grown into a six day course, that only covers about 50% of the subjects I tackled in the three day class. Why? Because it is no longer primarily a troubleshooting seminar. The vast majority of the class time, is now devoted to explaining how the equipment really functions and answering the following sorts of questions:

- Why do trays weep?
- Why do weeping trays have a low tray efficiency?
- What does tray efficiency actually mean anyway?
- Is there a way to design trays that do not weep?
- Why should an operator need to know why trays weep?
- Can a tray weep, even though the computer calculation says the tray cannot weep?

Several years ago, I began to make notes of the questions most frequently asked by my clients and students. Sometimes, it seems as if I have been asked and have responded to every conceivable process equipment question that could possibly be asked. Certainly, I have had plenty of practice in phrasing my answers, so that they are comprehensible to most process personnel, maintenance people and even management. We have tried to summarize these questions and answers in this book.

Like everybody else, I have answered questions without always being correct. But over the years, I have continued to learn. I have been taught by the source of all wisdom and knowledge - the process equipment itself. I am still learning. So you could say that this book is a kind of progress report of what I have learned so far. I like to think that my troubleshooting field work and revamp designs, have acted as a filter. This filter has and still is, removing from my store of knowledge, misconceptions as to the true nature of process equipment functions.

You do not need a technical degree to read and understand this text. Certainly this is a technical book. But the math and science discussed, is high school math and science. We have traded precision for simplicity in crafting this book.

Liz and I would be happy to discuss any questions you might have pertaining to the process equipment discussed in our book. You can phone or fax us in the United States at:

- phone: (504) 887-7714
- FAX: (504) 456-1835

But if you call us with a question, my first response is likely to be, "Have you looked the problem over in the field?"

Norman P. Lieberman

How Trays Work: Flooding

Downcomer Backup

Distillation towers are the heart of a process plant, and the working component of a distillation column is the tray. A tray consists of the following components, as shown in Fig. 1.1:

- Overflow, or outlet weir
- The downcomer
- The tray deck

There are two types of tray decks: perforated trays and bubble-cap trays. In this chapter, we describe only perforated trays, examples of which are

- Valves or flutter caps
- "V" grid, or extruded-valve caps
- Sieve decks
- "Jet" trays

Possibly 90 percent of the trays seen in the plant are of these types. Perforated tray decks all have one feature in common; they depend on the flow of vapor through the tray deck perforations, to prevent liquid from leaking through the tray deck. As we will see later, if liquid bypasses the outlet weir, and leaks through the tray deck onto the tray below, tray separation efficiency will suffer.

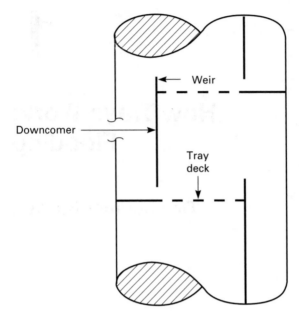

Figure 1.1 Perforated trays.

Tray Efficiency

Distillation trays in a fractionator operate between 10 and 90 percent efficiency. It is the process person's job to make them operate as close to 90 percent efficiency as possible. Calculating tray efficiency is simple. Compare the vapor temperature leaving a tray to the liquid temperature leaving the trays. For example, the efficiency of the tray shown in Fig. 1.2 is 100 percent. The efficiency of the tray in Fig. 1.3 is 0 percent.

How about the 10 trays shown in Fig. 1.4? Calculate their average efficiency (answer is 10 percent). As the vapor temperature rising from the top tray equals the liquid temperature draining from the bottom tray, the 10 trays are behaving as a single perfect tray with 100 percent efficiency. But as there are 10 trays, each tray, on average, acts like one-tenth of a perfect tray.

Poor tray efficiency is caused by one of two factors:

- Flooding
- Dumping

In this chapter, we discuss problems that contribute to tray deck flooding.

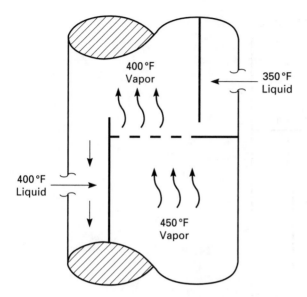

Figure 1.2 Hundred percent tray efficiency.

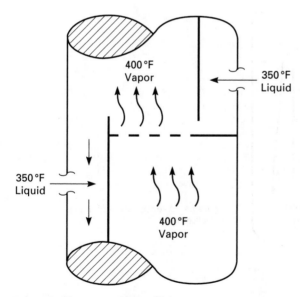

Figure 1.3 Zero percent tray efficiency.

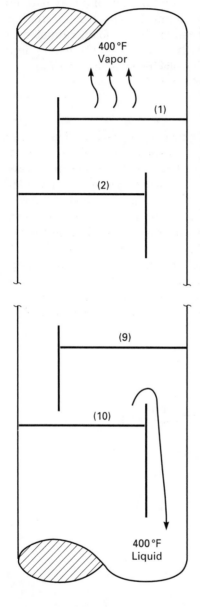

Figure 1.4 Average tray efficiency = 10 percent.

Downcomer Backup

Liquid flows across a tray deck toward the outlet weir. The liquid over-flows the weir, and drains through the downcomer, to the tray below.

Vapor bubbles up through the sieve holes, or valve caps, on the tray deck, where the vapor comes into intimate contact with the liquid. More precisely, the fluid on the tray is a froth or foam—that is, a mixture of vapor and liquid. In this sense, the function of a tray is to mix the vapor and liquid together, to form a foam. This foam separates back into a vapor and liquid in the downcomer. If the foam cannot drain quickly from a downcomer onto the tray below, then the foamy liquid or froth, will back up onto the tray above. This is called *flooding*.

Downcomer Clearance

Referring to Fig. 1.5, note that the downcomer is flooding. The cause is loss of the *downcomer seal*. The height of the outlet weir is below the bottom edge of the downcomer from the tray above. This permits vapor to flow up downcomer B. The upflowing vapor displaces the downflowing liquid. That is, the vapor pushes the liquid up onto the tray above—which is a cause of flooding. On the other hand, Fig. 1.6 shows what happens if the bottom edge of the downcomer is too close to the tray below. The high pressure drop needed for the liquid to escape from downcomer B onto tray deck 1 causes the liquid level in downcomer B to back up onto tray deck 2. Tray 2 then floods. Once tray 2 floods, downcomer C (shown in Fig. 1.6) will also back up and flood. This process will continue until all the tray decks and downcomers above downcomer B are flooded.

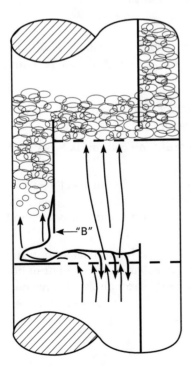

Figure 1.5 Flooding due to lack of a downcomer seal.

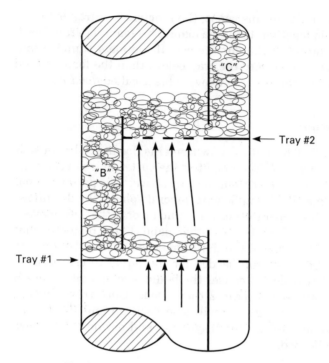

Figure 1.6 Flooding caused by inadequate downcomer clearance.

On the other hand, all trays in a tower below downcomer B will lose liquid levels and dry out, when flooding starts in downcomer B. Thus, the following rules apply:

- When flooding starts on a tray, all the trays above that point will also flood, but trays below that point will go dry.
- An early indication of flooding in a distillation column is loss of liquid level in the bottom of the column.
- If the downcomer clearance—which means the distance between the bottom edge of the downcomer and the tray below—is too great, the downcomer becomes unsealed. Vapor flows up the downcomer, and the trays above flood.
- If the downcomer clearance is too small, then liquid backs up in the downcomer, and the trays above flood. To calculate the height of liquid in the downcomer, due to liquid flowing through the downcomer clearance:

$$\Delta H = 0.6 \times V^2$$

where ΔH = inches of clear liquid backup in the downcomer, due to head loss under the downcomer

V = horizontal component of liquid velocity, in ft/s, as the liquid escapes from the downcomer

To guarantee a proper downcomer seal, the bottom edge of a downcomer should be about ½ inch below the top edge of the outlet weir. This dimension should be carefully checked by process personnel when a tower is opened for inspection. It is quite easy for sloppy tray installation to distort this critical factor.

Height of liquid on tray deck

As the liquid level on a tray increases, the height of liquid in the downcomer feeding this tray will increase by the same amount. Again, excessive downcomer liquid or froth levels result in flooding and loss of tray efficiency.

The liquid level on a tray is a function of two factors:

- The weir height
- The crest height

The weir height on many trays is adjustable. We usually adjust the weir height to between 2 and 3 inches. This produces a reasonable depth of liquid on the tray, to promote good vapor-liquid contact.

The crest height is similar to the height of water overflowing a dam. It is calculated from

$$\text{Crest height} = 0.4 \, (\text{GPM/inch (outlet) weir length})^{0.67}$$

where crest height = inches of clear liquid overflowing the weir

GPM = gallons (U.S.) per minute of liquid leaving the tray

The sum of the crest height plus the weir height equals the depth of liquid on the tray deck. One might now ask, "Is not the liquid level on the inlet side of the tray higher than the liquid level near the outlet weir?" While the answer is "Yes, water does flow downhill," we design the tray to make this factor small enough to neglect.

Vapor-Flow Pressure Drop

We have yet to discuss the most important factor in determining the height of liquid in the downcomer. This is the pressure drop of the vapor flowing through the tray deck. Typically, 50 percent of the level in the downcomer is due to the flow of vapor through the trays.

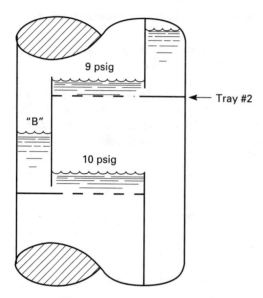

Figure 1.7 Vapor ΔP causes downcomer backup.

When vapor flows through a tray deck, the vapor velocity increases as the vapor flows through the small openings provided by the valve caps, or sieve holes. The energy to increase the vapor velocity comes from the pressure of the flowing vapor. A common example of this is the pressure drop we measure across an orifice plate. If we have a pipeline velocity of 2 ft/s and an orifice plate hole velocity of 40 ft/s, then the energy needed to accelerate the vapor as it flows through the orifice plate comes from the pressure drop of the vapor itself.

Let us assume that vapor flowing through a tray deck undergoes a pressure drop of 1 psig (lb/in² gauge). Figure 1.7 shows that the pressure below tray deck 2 is 10 psig and the pressure above tray deck 2 is 9 psig. How can the liquid in downcomer B flow from an area of low pressure (9 psig) to an area of high pressure (10 psig)? The answer is gravity, or liquid head pressure.

The height of water needed to exert a liquid head pressure of 1 psig is equal to 28 in of water. If we were working with gasoline, which has a specific gravity of 0.70, then the height of gasoline needed to exert a liquid head pressure of 1 psig would be 28 in/0.70 = 40 in of clear liquid.

Total height of liquid in downcomer

To summarize, the total height of clear liquid in the downcomer is the sum of four factors:

- Liquid escape velocity from the downcomer onto the tray below.
- Weir height.
- Crest height of liquid overflowing the outlet weir.
- The pressure drop of the vapor flowing through the tray above the downcomer. (Calculating this pressure drop is discussed in Chap. 2.)

Unfortunately, we do not have clear liquid, either in the downcomer, on the tray itself, or overflowing the weir. We actually have a froth or foam called *aerated liquid*. While the effect of this aeration on the specific gravity of the liquid is largely unknown and is a function of many complex factors (surface tension, dirt, tray design, etc.), an *aeration factor* of 50 percent is often used for many hydrocarbon services.

This means that if we calculated a clear liquid level of 12 inches in our downcomer, then we would actually have a foam level in the downcomer of 12 in/0.50 = 24 in of foam.

If the height of the downcomer, plus the height of the weir, were 24 in, then a downcomer foam height of 24 in would correspond to downcomer flooding. This is sometimes called *liquid flood*.

This discussion assumes that the cross-sectional area of the downcomer is adequate for reasonable vapor-liquid separation. If the downcomer loading (GPM/ft^2 of downcomer top area) is less than 150, this assumption is okay, at least for most clean services.

Jet Flood

Figure 1.8 is a realistic picture as to what we would see if our towers were made of glass. In addition to the downcomers and tray decks containing froth or foam, there is a quantity of spray, or entrained liquid, lifted above the froth level on the tray deck. The force that generates this entrainment is the flow of vapor through the tower. The spray height of this entrained liquid is a function of two factors:

- The foam height on the tray
- The vapor velocity through the tray

High vapor velocities, combined with high foam levels, will cause the spray height to hit the underside of the tray above. This causes mixing of the liquid from a lower tray, with the liquid on the upper tray. This backmixing of liquid reduces the separation, or tray efficiency, of a distillation tower.

When the vapor flow through a tray increases, the height of froth in the downcomer draining the tray will also increase. This does not affect

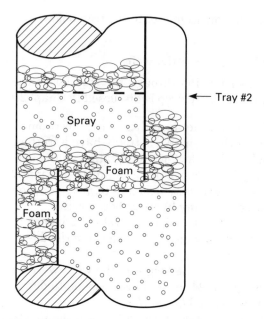

Figure 1.8 Entrainment causes jet flood.

the foam height on the tray deck until the downcomer fills with foam. Then, a further increase in vapor flow causes a noticeable increase in the foam height of the tray deck, which then increases the spray height.

When the spray height from the tray below hits the tray above, this is called the *incipient flood point,* or the initiation of jet flooding. Note, though, that jet flood may be caused by excessive downcomer backup. It is simple to see, in a glass column separating colored water from clear methanol, how tray separation efficiency is reduced as soon as the spray height equals the tray spacing. And while this observation of the onset of incipient flood is straightforward in a transparent tower, how do we observe the incipient flooding point in a commercial distillation tower?

Incipient Flood

A fundamental concept

Figure 1.9 illustrates the operation of a simple propane-butane splitter. The tower controls are such that both the pressure and bottoms temperature are held constant. This means that the percent of propane in the butane bottoms product is held constant. If the operator increases the top reflux flow, here is what will happen:

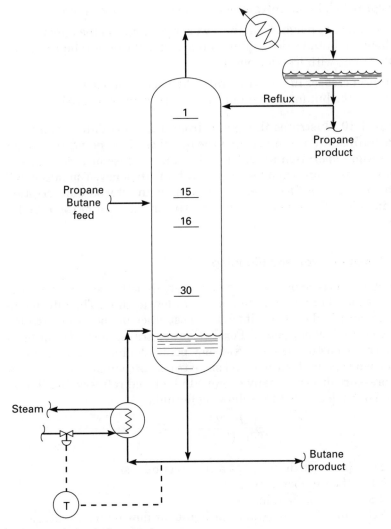

Figure 1.9 A simple depropanizer.

- The tower-top temperature drops
- The amount of butane in the overhead propane product drops
- The tower-bottom temperature starts to fall
- The reboiler duty increases, to restore the tower-bottom temperature to its set point
- The weight flow of the vapor, and the vapor velocity through the trays, increase

- The spray height, or entrainment, between the trays, increases
- When the spray height from the lower trays, impacts the upper trays, the heavier, butane-rich liquid contaminates the lighter liquid on the upper trays, with heavier butane
- Further increases in the reflux rate, then act to increase, rather than decrease, the butane content of the overhead propane product

Figure 1.10 illustrates this point, from plant test data obtained in a Texas refinery. Point A is called the *incipient flood point,* that point in the towers operation at which either an increase or a decrease in the reflux rate results in a loss of separation efficiency. You might call this the *optimum reflux rate;* that would be an alternate description of the incipient flood point, neglecting the energy cost of the reboiler steam.

Tower Pressure Drop and Flooding

It is a characteristic of process equipment, that the best operation is reached, at neither a very high nor a very low loading. The intermediate equipment load that results in the most efficient operation is called the "the best efficiency point." For distillation trays, the incipient flood point corresponds to the best efficiency point. We have correlated this best efficiency point, for valve and sieve trays, as compared to the measured pressure drops in many chemical plant and refinery distillation towers. We have derived the following formula:

$$\frac{(\Delta P)\,(28)}{(NT)\,(TS)\,(SG)} = K$$

where ΔP = pressure drop across a tray section, psi
  NT = the number of trays
  TS = tray spacing, inches
  SG = specific gravity of clear liquid, at flowing temperatures

On the basis of hundreds of field measurements, we have observed

K = 0.18 to 0.25; tray operation close to its best efficiency point
K = 0.35 to 0.40; tray suffering from entrainment—increase in reflux rate, noticeably reduces tray efficiency
K = ≥0.5; tray is in fully developed flood—opening a vent on the overhead vapor line will blow out liquid, with the vapor
K = 0.10 to 0.12; tray deck is suffering from low tray efficiency, due to tray deck leaking
K = 0.00; the liquid level on the tray is zero, and quite likely the trays are lying in the bottom of the column

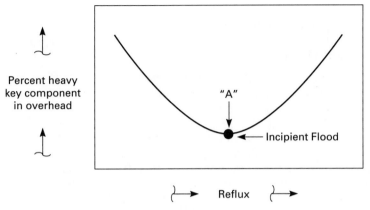

Figure 1.10 Definition of the incipient flood concept.

Carbon steel trays

One of the most frequent causes of flooding is the use of carbon steel trays. Especially when the valve caps are also carbon steel, the valves have a tendency to stick in a partially closed position. This raises the pressure drop of the vapor flowing through the valves, which, in turn, pushes up the liquid level in the downcomer draining the tray. The liquid can then back up onto the tray deck, and promote jet flood, due to entrainment.

Of course, any factor (dirt, polymers, gums, salts) that causes a reduction in the open area of the tray deck will also promote jet flooding. Indeed, most trays flood below their calculated flood point, because of these sorts of problems. Trays, like people, rarely perform quite up to expectations.

2

How Trays Work: Dumping

Weeping through Tray Decks

A distillation tray works efficiently, when the vapor and liquid come into intimate contact on the tray deck. To this end, the liquid should flow evenly across the tray deck. The vapor should bubble up evenly through the perforations on the tray deck. The purpose of the outlet weir is to accomplish both these objectives, as follows:

1. Uneven liquid flow across the tray deck is particularly detrimental to good vapor-liquid mixing. For example, if half of the tray deck has stagnant liquid, then the vapor bubbling through the stagnant liquid cannot alter its composition. Uneven liquid flow is promoted by the outlet weir being out-of-level. Liquid will tend to flow across that portion of the tray, with a lower-than-average weir height. The portion of the tray upstream of the high part of the outlet weir will contain stagnant liquid. However, if the crest height (i.e., the height of liquid over the weir) is large, compared to the out-of-levelness of the tray, then an even liquid flow across the tray will result. To achieve a reasonable crest height above the outlet weir, a *weir loading* of at least 2 GPM per inch of weir length is needed. When liquid flows are small, the tray designer employs a picket weir, as shown in Fig. 2.1.

2. Uneven vapor flow bubbling-up through the tray deck will promote vapor-liquid channeling. This sort of channeling accounts for many trays that fail to fractionate up to expectations. To understand the cause of this channeling, we will have to quantify total tray pressure drop.

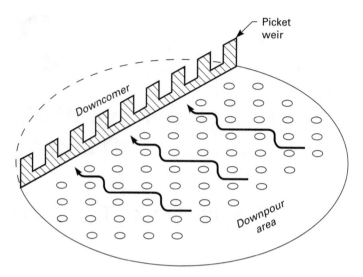

Figure 2.1   Picket weir promotes even liquid cross-flow, at low flows.

Tray Pressure Drop

Total tray ΔP

Figure 2.2 shows a simple sieve tray, with a single hole. Why is it that the liquid flows over the 3-in outlet weir, rather than simply draining down through the sieve hole? It is the force of the vapor (or better, the velocity of the vapor), passing through the sieve hole, which prevents

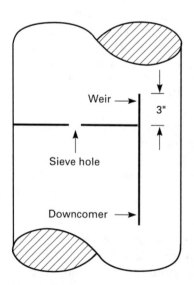

Figure 2.2 A simplified sieve tray.

the liquid from draining down the sieve hole. This is true whether we are dealing with a valve cap, extruded perforation, or a sieve hole. The valve cap does not act as a check valve, to keep liquid on the tray. The author's industrial experience has proved this unpleasant fact on numerous occasions.

On the other hand, bubble caps (or even the more ancient tunnel cap trays) are different, in that they do not depend on the vapor flow to retain the liquid level on the tray deck. More on this later. For now, just recall that we are dealing only with perforated tray decks.

Dry-tray pressure drop

For the force of the upflowing vapor to stop the liquid from leaking through the sieve hole shown in Fig. 2.2, the pressure drop of the vapor flowing through the hole has to equal the weight of liquid on the tray deck. The weight of liquid trying to force its way through the sieve hole is proportional to the depth of liquid on the tray deck. The pressure drop of the vapor, as it accelerates through the sieve hole, is

$$\Delta P_{dry} = K \frac{D_V}{D_L} Vg^2$$

where ΔP_{dry} = dry tray pressure drop, in inches of clear liquid
$\quad\quad D_V$ = density of vapor, lb/ft^3
$\quad\quad D_L$ = density of liquid, lb/ft^3
$\quad\quad Vg$ = velocity of vapor or gas flowing through the sieve hole, ft/s
$\quad\quad K$ = an orifice coefficient, which can be as low as 0.3 for a smooth hole in a thick plate and 0.6 to 0.95 for various valve tray caps

Hydraulic tray pressure drop

The weight of liquid on a tray is created by the weir height plus the crest height. We have defined the crest height (in inches of clear liquid) in Chap. 1, as

Crest height = 0.4 (GPM/inch outlet weir length)$^{0.67}$

The actual height of fluid overflowing the weir is quite a bit greater than we calculate with this formula. The reason is that the fluid over-flowing the weir is not clear liquid, but aerated liquid—that is, foam. The fluid on the tray deck, below the top of the weir, is also foam. This reduces the effective weight of the liquid on the tray due to aeration. To summarize, the weight of liquid on the tray, called the *hydraulic tray pressure drop,* is

$$\Delta P_{hyd} = AF \times WH + 0.4 \text{ (GPM/inch outlet weir length)}^{0.67}$$

where ΔP_{hyd} = hydraulic tray pressure drop, in inches of clear liquid
    WH = weir height, in
    AF = aeration factor
    GPM = gallons (U.S.) per minute

The aeration factor AF is the relative density of the foam, to the density of the clear liquid. It is a combination of complex factors, but is typically 0.5.

Calculated total tray pressure drop

The sum of the dry tray pressure drop (ΔP_{dry}) plus the hydraulic tray pressure drop (ΔP_{hyd}) equals the total tray pressure drop (ΔP_{total}):

$$\Delta P_{total} = \Delta P_{dry} + \Delta P_{hyd}$$

expressed in inches of clear liquid

When the dry tray pressure drop is significantly less than the hydraulic tray pressure drop, then the tray will start to leak or weep, and tray efficiency will be adversely affected.

When the dry tray pressure drop is significantly greater than the hydraulic tray pressure drop, then the liquid on the tray can blow off of the tray deck, and tray efficiency will be adversely affected.

For a tray to function reasonably close to its best efficiency point, the dry tray pressure drop must be roughly equal (± 50 percent) to the hydraulic tray pressure drop:

$$\Delta P_{dry} = \Delta P_{hyd}$$

This concept is the basis for tray design for perforated tray decks. While various valve tray vendors maintain that this rule does not hold for their equipment, it is the author's industrial experience that valve trays leak just as badly as do sieve trays, at low vapor hole velocities. To summarize:

$$\Delta P_{total} = K \frac{D_V}{D_L} Vg^2 + AF \times WH + 0.4 \text{ (GPM/inch outlet weir length)}^{0.67}$$

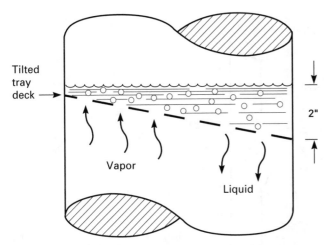

Figure 2.3   Out-of-level tray causing vapor-liquid channeling.

Other Causes of Tray Inefficiency

Out-of-level trays

When trays weep, efficiency may not be significantly reduced. After all, the dripping liquid will still come into good contact with the upflowing vapor. But this statement would be valid only if the tray decks were absolutely level. And in the real world, especially in large (>6-ft)-diameter columns, there is no such thing as a "level" tray. Figure 2.3 shows the edge view of a tray that is 2 inches out-of-level.

As illustrated, liquid accumulates on the low side of this tray. Vapor, taking the path of least resistance, preferentially bubbles up through the high side of the tray deck. To prevent liquid from leaking through the low side of the tray, the dry tray pressure drop must equal or exceed the sum of the weight of the aerated liquid retained on the tray by the weir *plus* the crest height of liquid over the weir *plus* the 2-in out-of-levelness of the tray deck.

Once the weight of liquid on one portion—the lowest area—of a tray deck exceeds the dry tray pressure drop, the hydraulic balance of the entire tray is ruined. Vapor flow through the low area of the tray deck ceases. The aeration of the liquid retained by the weir on the tray deck stops, and hence the hydraulic tray pressure drop increases even more. As shown in Fig. 2.3, the liquid now drains largely through the low area of the tray. The vapor flow bubbles mainly through the higher area of the tray deck. This phenomenon is termed vapor-liquid *channeling*. Channeling is the primary reason for reduced distillation tray efficiency.

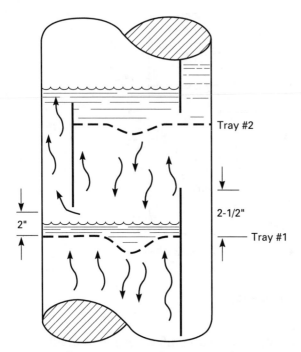

Figure 2.4 Sagging tray ruins downcomer seal.

The common reason for out-of-levelness of trays is sagging of the tray decks. Sags are caused by pressure surges and sloppy installation. Sometimes, the tray support rings might not be installed level; or the tower itself might be out-of-plumb (meaning the tower itself may not be truly vertical).

Loss of downcomer seal

We stated in Chap. 1 that the top edge of the outlet weir is maintained about ½ in above the bottom edge of the inlet downcomer, to prevent vapor from flowing up the downcomer. This is called a *½-in positive downcomer seal*. But for this seal to be effective, the liquid must overflow the weir. If all the liquid is weeping through the tray deck, then there will be no flow over the weir, and the height of the weir will become irrelevant. Figure 2.4 shows the result of severe tray deck leakage:

- The downcomer seal is lost on tray deck 1.
- Vapor flows up the downcomer, between tray decks 1 and 2.

- The dry tray pressure drop through tray 2 decreases, due to low vapor flow through the tray deck.

- The hydraulic tray pressure drop on tray 2 increases, due to increased liquid level.

- Tray 2 will now start to weep, with the weeping concentrated on the low area of the tray.

- Tray 2 now has most of its vapor feed flowing up through its outlet downcomer, rather than the tray deck; and most of its liquid flow, leaking through its tray deck.

The net result of this unpleasant scenario is loss of both vapor-liquid contacting and tray efficiency. Note how the mechanical problems (i.e., levelness) of tray 1 ruins the tray efficiency of both trays 1 and 2.

Bubble-cap Trays

The first continuous distillation tower built, was the "patent still" used in Britain to produce Scotch whiskey, in 1835. The patent still is, to this day, employed to make apple brandy, in southern England. The original still, and the one I saw in England in 1992, had ordinary bubble-cap trays (except downpipes instead of downcomers, were used). The major advantage of a bubble-cap tray is that the tray deck is leakproof. As shown in Fig. 2.5, the riser inside the cap is above the top of the out-

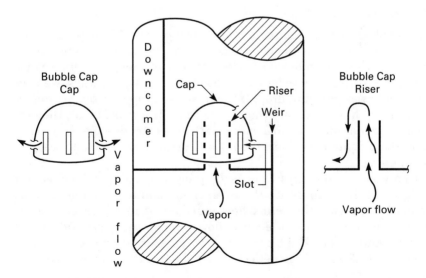

Figure 2.5   A bubble cap showing vapor pathway in operation.

major advantage of a bubble-cap tray is that the tray deck is leakproof. As shown in Fig. 2.5, the riser inside the cap is above the top of the outlet weir. This creates a mechanical seal on the tray deck, which prevents liquid weeping, regardless of the vapor flow.

Bubble-cap trays may be operated over a far wider range of vapor flows, without loss of tray efficiency. It is the author's experience that bubble-cap trays fractionate better in commercial service than do perforated (valve, or sieve) trays. Why, then, are bubble-cap trays rarely used in a modern distillation tower?

There really is no proper answer to this question. It is quite likely that the archaic, massively thick, bolted-up, cast-iron bubble-cap or tunnel-cap tray was the best tray ever built. However, compared to a modern valve tray, bubble-cap trays

- Were difficult to install, because of their weight.

- Have about 15 percent less capacity because, when vapor escapes from the slots on the bubble cap, it is moving in a horizontal direction. The vapor flow must turn 90°. This change of direction promotes entrainment and, hence, jet flooding.

- Are more expensive to purchase.

But in the natural-gas fields, where modern design techniques have been slow to penetrate, bubble-cap trays are still widely employed, to dehydrate and sweeten natural gas in remote locations.

Distillation tower turndown

The problem we have been discussing—loss of tray efficiency due to low vapor velocity—is commonly called *turndown*. It is the opposite of flooding, which is indicated by loss of tray efficiency, at high vapor velocity. To discriminate between flooding and weeping trays, we measure the tower pressure drop. If the pressure drop per tray, expressed in inches of liquid, is more than three times the weir height, then the poor fractionation is due to flooding. If the pressure drop per tray is less than the height of the weir, then poor fractionation is due to weeping or dumping.

One way to stop trays from leaking or weeping is to increase the reflux rate. Assuming that the reboiler is on automatic temperature control, increasing the reflux flow must result in increased reboiler duty. This will increase the vapor flow through the trays and the dry tray pressure drop. The higher dry tray pressure drop may then stop tray deck leakage. The net effect is that the higher reflux rate restores the tray efficiency.

largest operating cost for many process units is the energy supplied to the reboilers. We should therefore avoid high reflux rates, and try to achieve the best efficiency point for distillation tower trays, at a minimum vapor flow. This is best done by designing and installing the tray decks and outlet weirs as level as possible. Damaged tray decks should not be reused, unless they can be restored to their proper state of levelness.

3

Why Control
Tower Pressure

Options for
Optimizing Tower
Operating Pressure

While we realize that distillation towers are designed with a control scheme to fix the tower pressure, why is this necessary?

Naturally, we do not want to overpressure the tower, and pop open the safety relief valve. Alternatively, if the tower pressure gets too low, we could not condense the reflux. Then, the liquid level in the reflux drum would fall and the reflux pump would lose suction and cavitate. But assuming that we have plenty of condensing capacity, and are operating well below the relief valve set pressure, why do we attempt to fix the tower pressure? Further, how do we know what pressure target to select?

I well remember one pentane-hexane splitter in Toronto. The tower simply could not make a decent split, regardless of the feed or reflux rate selected. The tower-top pressure was swinging between 12 and 20 psig. The flooded condenser pressure control valve, shown in Fig. 3.1, was operating between 5 and 15 percent open, and hence it was responding in a nonlinear fashion (most control valves work properly only at 20 to 75 percent open). The problem may be explained as follows.

The liquid on the tray deck was at its bubble, or boiling point. A sudden decrease in the tower pressure caused the liquid to boil violently. The resulting surge in vapor flow promoted jet entrainment, or flooding.

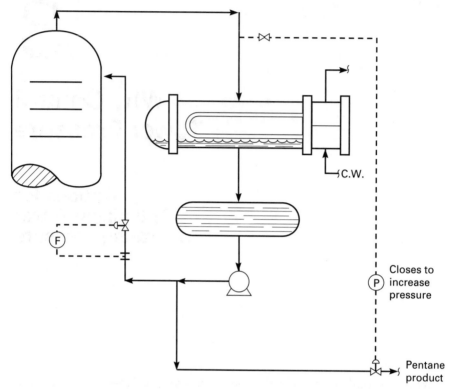

Figure 3.1 Flooded condenser pressure control.

Alternately, the vapor flowing between trays was at its dew point. A sudden increase in tower pressure caused a rapid condensation of this vapor and a loss in vapor velocity through the tray deck holes. The resulting loss in vapor flow caused the tray decks to dump.

Either way, erratic tower pressure results in alternating flooding and dumping, and therefore reduced tray efficiency. While gradual swings in pressure are quite acceptable, no tower can be expected to make a decent split with a rapidly fluctuating pressure.

Selecting an Optimum Tower Pressure

The process design engineer selects the tower design operating pressure as follows:

- Determines the maximum cooling water or ambient air temperature that is typically expected on a hot summer day in the locale where the plant is to be built.

- Next calculates the condenser outlet, or reflux drum temperature, that would result from the above water or air temperature (as discussed in Chap. 13).

- Referring to Fig. 3.2, the designer next calculates the pressure in the reflux drum, assuming that the condensed liquid is at its bubble point (as discussed in Chap. 8). Adding 5 or 10 psig to this pressure, for pressure loss in the overhead condenser and associated piping, the designer then determines the tower-top pressure.

Of course, the unit operator can physically deviate from this design pressure; but to what purpose?

Raising the Tower Pressure Target

I once had a contract in a Denver refinery to revamp a hydrocracker fractionator, which produced naphtha, jet fuel, and diesel. The bottleneck was tray flooding. At higher feed rates, the kerosene would carry over into the overhead naphtha product. My initial plant inspection showed that the tower-top pressure was 24 psig. The relief valve was set at 50 psig. By raising the tower operating pressure to 30 psig, the flooding was stopped and my contract disappeared. Why?

Ambient pressure in Denver is about 13 psia (lb/in² absolute) (vs. 14.7 psia at sea level). Higher pressures reduce the volumetric flow of vapor. In other words, volume is inversely proportional to pressure:

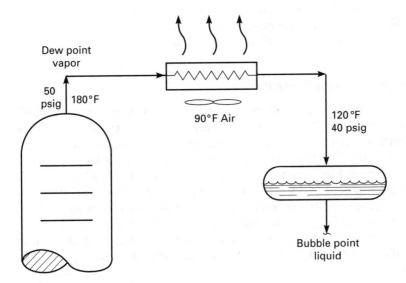

Figure 3.2 Calculating the tower design pressure.

$$\text{Volume} \sim \frac{1}{\text{pressure}}$$

The pressure we are concerned with is the absolute pressure:

$$\text{Initial pressure} = 24 \text{ psig} + 13 \text{ psi} = 37 \text{ psia}$$

$$\text{Final pressure} = 30 \text{ psig} + 13 \text{ psi} = 43 \text{ psia}$$

The absolute tower pressure (in psia) increased by 17 percent, and hence the volume (as well as the velocity of vapor through the valve tray caps) declined by 17 percent. The reduced vapor velocity reduced the dry tray pressure drop, thus reducing both the spray height above the tray deck and the liquid backup in the downcomers.

Another reason to raise tower pressure is to permit higher reflux rates. If the pressure controller in Fig. 3.1 is set too low, then during hot weather, when condenser capacity becomes marginal, the level in the reflux drum will be lost. If we then raise the pressure set point, the drum will refill—but why?

Raising the tower pressure also increases the reflux drum pressure, raising, in turn, the temperature at which the vapors condense. The rate of condensation is then calculated from the following:

$$Q = U \times A \, (T_C - T_A)$$

where Q = rate of condensation, Btu/h
A = heat-exchanger surface area, ft^2
U = heat-transfer coefficient, Btu/[(h)(ft^2)(°F)]
T_C = condensation temperature of vapors, °F
T_A = temperature of air or cooling water, °F

Lowering the Tower Pressure

In general, distillation columns should be operated at a low pressure. For example, Fig. 3.3 shows an isobutane-normal butane stripper. This fractionator is performing poorly. A computer simulation of the column has been built. The column has 50 actual trays. But in order to force the computer model to match existing operating parameters (reflux rate, product compositions), 10 theoretical separation stages (i.e., 10 trays, each 100 percent efficient) must be used in the model. This means that the trays are developing an actual tray efficiency of only 20 percent.

A field measurement indicated a pressure drop of 2.0 psi. Assuming a specific gravity of 0.50, then the pressure drop per tray, in inches of liquid is:

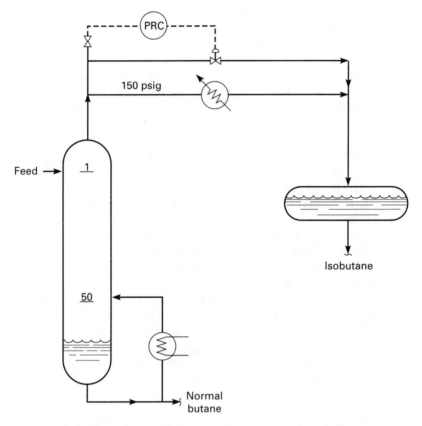

Figure 3.3 Isobutane stripper with hot-vapor bypass pressure controller.

$$\frac{2.0 \text{ psi tower } \Delta P}{50 \text{ tray}} \times \frac{28 \text{ in H}_2\text{O}}{1 \text{ psi}} \times \frac{1.0 \text{ (SG water)}}{0.50 \text{ (SG butane)}}$$

$$= 2.24 \text{ in liquid } \Delta P \text{ per tray}$$

As the weir height of the trays is 3 in, it is a safe assumption that the low tray efficiency is due to tray deck dumping, rather than flooding. As shown in Fig. 3.3, this column has no reflux. This is a typical design for strippers; when feed is introduced on the top tray, there is no need for reflux.

In order to improve tray efficiency, it will be necessary to increase the vapor velocity through the trays, so as to increase the pressure drop to at least 4 or 5 in of liquid per tray. If the reboiler duty were simply increased, the concentration of the heavy component—normal butane—in the light overhead product—isobutane—would escalate exponentially. Another method, however, that does not involve increasing either the reboiler duty or the *mass flow* of vapor through the trays, can be used to increase vapor velocity.

By lowering the tower operating pressure, the volume of vapor flow bubbling up through the tray decks may be increased, without changing the mass flow. For instance, if the tower pressure were reduced from 105 psig (or 120 psia) to 45 psig (or 60 psia), then the velocity of vapor through the sieve holes on the trays would double. This would lead to a substantial increase in the dry tray pressure drop, and hence reduce tray deck leakage.

To lower the tower pressure, the hot-vapor bypass pressure recorder controller (PRC) valve is closed. This forces more vapor through the condenser, which, in turn, lowers the temperature in the reflux drum. As the liquid in the reflux drum is at its bubble point, reducing the reflux drum temperature will reduce the reflux drum pressure. As the stripper tower pressure floats on the reflux drum pressure, the pressure in the tower will also decline.

The net effect of reducing the stripper pressure was to greatly reduce the amount of isobutane in the heavier normal butane bottoms product. Undoubtedly, most of the improvement in fractionation was due to enhanced tray efficiency, which resulted from suppressing tray deck leaking, or dumping. But there was a secondary benefit of reducing tower pressure: increased *relative volatility*.

Relative volatility

The chart shown in Fig. 3.4, is called a *Cox,* or *vapor-pressure,* chart. It shows the pressure developed by pure-component liquids, at various temperatures. The interesting aspect of this chart is that the sloped

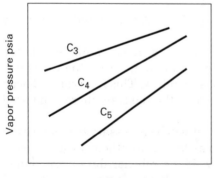

Figure 3.4 Vapor-pressure chart.

lines, representing the vapor pressures of pure hydrocarbon components, spread apart at lower pressures. This results in an increase in the ratio of the vapor pressures of any two components.

The vapor pressure of a light component at a given temperature, divided by the vapor pressure of a heavier component at the same temperature, is called the *relative volatility*. For practice, calculate the relative volatility of isobutanes and normal pentane at 140°F (answer: 4.0). Next, calculate their relative volatility at 110°F (answer: 4.9).[1]

Note that the relative volatility has increased by about 20 percent at the lower temperature and pressure. This increase in relative volatility allows one to make a better split at a given reflux rate, or to make the same split at a lower reflux rate. We can quantify this last statement as follows:

$$\frac{(RVH - 1) - (RVL - 1)}{(RVH - 1)} = DRF$$

where RVH = relative volatility at a high pressure
RVL = relative volatility at a low pressure
DRF = percent reduction in the reflux rate, when the same degree of fractionation is desired

Reducing reflux saves reboiler duty. Also, the lower pressure will reduce the tower-bottom temperature, and this also cuts the reboiler energy requirement. For most distillation towers, the energy cost of the reboiler duty is the main component of the total operating cost to run the tower.

Incipient flood point

As an operator reduces the tower pressure, three effects occur simultaneously:

- The relative volatility increases.
- Tray deck leakage decreases.
- Entrainment, or spray height, increases.

The first two factors help make fractionation better, the last factor makes fractionation worse. How can an operator select the optimum tower pressure, to maximize the benefits of enhanced relative volatility, and reduced tray deck dumping, without unduly promoting jet flooding due to entrainment?

To answer this fundamental question, we should realize that reducing the tower pressure will also reduce both the tower-top temperature

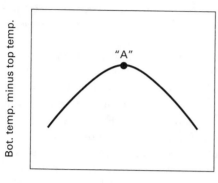

Bot. temp. minus top temp.

Tower pressure

Figure 3.5 Point "A" represents the optimum tower pressure.

and the tower-bottom temperature. So the change in these temperatures, by themselves, is not particularly informative. But if we look at the difference between the bottom and top temperatures, this difference is an excellent indication of fractionation efficiency. The bigger this temperature difference, the better the split. For instance, if the tower-top and tower-bottom temperatures are the same for a 25-tray tower, what is the average tray efficiency? (Answer: 100 percent ÷ 25 = 4 percent.)

Figure 3.5 illustrates this relationship. Point A is the incipient flood point. In this case, the *incipient flood point* is defined as that operating pressure that maximizes the temperature difference across the tower at a particular reflux rate. How, then, do we select the optimum tower pressure, to obtain the best efficiency point for the trays? Answer—look at the temperature profile across the column.

The Phase Rule in Distillation

This is, perhaps, an idea you remember from high school, but never quite understood. The phase rule corresponds to determining how many independent variables we can fix in a process, before all the other variables become *dependent variables.* In a reflux drum, we can fix the temperature and composition of the liquid in the drum. The temperature and composition are called *independent variables.* The pressure in the drum could now be calculated from the chart in Fig. 3.4. The pressure is a dependent variable. The *phase rule* for the reflux drum system states that we can select any two variables arbitrarily (temperature, pressure, or composition), but then the remaining variable is fixed.

A simple distillation tower, like that shown in Fig. 3.2, also must obey its own phase rule. Here, because the distillation tower is a more complex system than the reflux drum, there are three independent variables that must be specified. The operator can choose from a large number of variables, but must select no more than three from the following list:

- Tower pressure
- Reflux rate, or reflux ratio
- Reboiler duty
- Tower-top temperature
- Tower-bottom temperature
- Overhead product rate
- Bottoms product rate
- Overhead product composition
- Bottoms product composition

The prior discussion assumes that the feed rate, feed composition, and heat content (enthalpy) are fixed. My purpose in presenting this review of the phase rule is to encourage the routine manipulation of tower operating pressures, in the same sense, and with the same objectives, as adjusting reflux rates. Operators who arbitrarily run a column, at a fixed tower pressure, discards one-third of the flexibility available to them, to operate the column in the most efficient fashion. And this is true, regardless of whether the objective is to save energy or improve the product split.

Reference

1. API Data Book, Vapor Pressure Section, 1986 edition.

4

What Drives Distillation Towers

Reboiler Function

An internal-combustion engine drives a car. Pumps are driven by turbines or motors. Jet planes are pushed by the thrust of an axial compressor.

The Reboiler

All machines have drivers. A distillation column is also a machine, driven by a *reboiler*. It is the heat duty of the reboiler, supplemented by the heat content (enthalpy) of the feed, that provides the energy to make a split between light and heavy components. A useful example of the importance of the reboiler in distillation comes from the venerable use of sugar cane, in my home state of Louisiana.

If the cut cane is left in the fields for a few months, its sugar content ferments to alcohol. Squeezing the cane then produces a rather low-proof alcoholic drink. Of course, one would naturally wish to concentrate the alcohol content by distillation, in the still shown in Fig. 4.1.

The alcohol is called the "light" component, because it boils at a lower temperature then water; the water is called the "heavy" component, because it boils at a higher temperature then alcohol. Raising the top reflux rate will lower the tower-top temperature, and reduce the amount of the heavier component, water, in the overhead alcohol product. But what happens to the weight of vapor flowing up through the trays? Does the flow go up, go down, or remain the same?

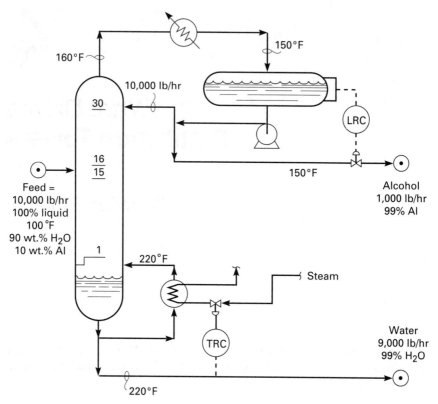

Figure 4.1 Alcohol-water splitter.

There are two ways to answer this question. Let's first look at the reboiler. As the tower-top temperature shown in Fig. 4.1 goes down, more of the lighter, lower-boiling-point alcohol is refluxed down the tower. The tower-bottom temperature begins to drop, and the steam flow to the reboiler is automatically increased by the action of the temperature recorder controller (TRC). As the steam flow to the reboiler increases, so does the reboiler duty (or energy injected into the tower in the form of heat). Almost all the reboiler heat or duty is converted to vaporization. We will prove this statement mathematically later in this chapter. The increased vapor leaving the reboiler then bubbles up through the trays, and hence the flow of vapor is seen to increase, as the reflux rate is raised.

Now let's look at the reflux drum. The incremental reflux flow comes from this drum. But the liquid in this drum comes from the condenser. The feed to the condenser is vapor from the top of the tower. Hence, as we increase the reflux flow, the vapor rate from the top of the tower must increase. One way of summarizing these results is to say that the *reflux comes from the reboiler.*

The statement that the mass, or weight flow of vapor through the trays, increases as the refluxed rate is raised is based on the reboiler being on automatic temperature control. If the reboiler were on manual control, then the flow of steam and the reboiler heat duty would remain constant as the reflux rate was increased, and the weight flow of vapor up the tower would remain constant as the top reflux rate was increased. But the liquid level in the reflux drum would begin to drop. The reflux drum level recorder controller (LRC) would close off to catch the falling level, and the overhead product rate would drop, in proportion to the increase in reflux rate. We can now draw some conclusions from the foregoing discussion:

- The flow of vapor leaving the top tray of the tower is equal to the flow of reflux, plus the flow of the alcohol overhead product.
- The overhead condenser heat-removal duty is proportional to the reboiler heat duty.
- The weight flow of vapor in a tower is controlled by one factor, and one factor only: *heat.*

An increase in reflux rate, assuming that the reboiler is on automatic temperature control, increases both the tray weir loading and the vapor velocity through the tray deck. This increases both the total tray pressure drop and the height of liquid in the tray's downcomer. Increasing reflux rates, with the reboiler on automatic temperature control, then will always push the tray closer to, or even beyond, the point of incipient flood.

Heat-Balance Calculations

If you have read this far, and understood what you have read, you will readily understand the following calculation. It is simply a repetition, with numbers, of the discussion previously presented. However, you will require the following values to perform the calculations:

- Latent heat of condensation of alcohol vapors = 400 Btu/lb
- Latent heat of condensation of water vapors = 1000 Btu/lb
- Specific heat of alcohol (vapor or liquid) = 0.6 Btu/[(lb)(°F)]
- Specific heat of water = 1.0 Btu/[(lb)(°F)]

The term *specific heat* refers to the sensible-heat content of either vapor or liquid. The specific heat is the amount of heat needed to raise the temperature on one pound of the vapor or liquid by 1°F. The term *latent heat* refers to the heat of vaporization, or the heat of condensation, needed to vaporize or condense one pound of liquid or vapor at constant temperature. Note that the heat of condensation is equal to the heat of vaporization. Each is referred to as the *latent heat*. The sum of the sensible heat, plus the latent heat, is called the *total heat content*, or *enthalpy*.

Returning to our example in Fig. 4.1, we wish first to determine the reboiler duty. To do this, we have to supply three heat requirements:

A. Heat 9000 lb/h of water from the 100°F feed temperature to the tower-bottom temperature of 220°F.

B. Heat 1000 lb/h of alcohol from the 100°F feed temperature (where the alcohol is a liquid) to the tower overhead temperature of 160°F (where the alcohol is a vapor).

C. Vaporize 10,000 lb/h of reflux from the 150°F reflux drum temperature to the tower overhead temperature of 160°F.

Solution to step A:

$$9000 \text{ lb/h} \times 1.0 \text{ Btu/[(lb)(°F)]} \times (220°F - 100°F) = 1{,}080{,}000 \text{ Btu/h}$$

Solution to step B:

$$1000 \text{ lb/h} \times 0.6 \text{ Btu/[(lb)(°F)]} \times (160°F - 100°F) + 1000 \text{ lb/h}$$
$$\times 400 \text{ Btu/lb} = 36{,}000 \text{ Btu/h} + 400{,}000 \text{ Btu/h} = 436{,}000 \text{ Btu/h}$$

Solution to step C:

$$10{,}000 \text{ lb/h} \times 0.6 \text{ Btu/[(lb)(°F)]} \times (160°F - 150°F)$$
$$+ 10{,}000 \text{ lb/h} \times 400 \text{ Btu/lb}$$
$$= 60{,}000 \text{ Btu/h} + 4{,}000{,}000 \text{ Btu/h} = 4{,}060{,}000 \text{ Btu/h}$$

The reboiler duty is then the sum of A + B + C = 5,576,000 Btu/h.

The next part of the problem is to determine the vapor flow to the bottom tray. If we assume that the vapor leaving the reboiler is essentially steam, then the latent heat of condensation of this vapor is 1000 Btu/lb. Hence the flow of vapor (all steam) to the bottom tray is

$$5{,}576{,}000 \text{ Btu/h} \div 1000 \text{ Btu/lb} = 5576 \text{ lb/h}$$

What about the vapor flow leaving the top tray of our splitter? That is simply the sum of the reflux plus the overhead product:

$$10,000 \text{ lb/h} + 1000 \text{ lb/h} = 11,000 \text{ lb/h}$$

How about the condenser duty? That is calculated as follows:

$$11,000 \text{ lb/h} \times 0.6 \text{ Btu/[(lb)(°F)]} \times (160°F - 150°F) + 11,000 \text{ lb/h}$$
$$\times 400 \text{ Btu/lb} = 66,000 \text{ Btu/h} + 4,400,000 \text{ Btu/h} = 4,466,000 \text{ Btu/h}$$

We can draw the following conclusions from this example:

- The condenser duty is usually a little smaller than the reboiler duty.
- Most of the reboiler heat duty usually goes into generating reflux.
- The flow of vapor up the tower is created by the reboiler.

For other applications, these statements may be less appropriate. This is especially so when the reflux rate is much smaller than the feed rate. But if you can grasp these calculations, then you can appreciate the concept of the reboiler, acting as the engine to drive the distillation column.

Effect of feed preheat

Up to this point, we have suggested that the weight flow of vapor up the tower is a function of the reboiler duty only. Certainly, this cannot be completely true. If we look at Fig. 4.2, it certainly seems that increasing the heat duty on the *feed preheater* will reduce the reboiler duty.

Let us assume that both the reflux rate and the overhead propane product rate are constant. This means that the total heat flow into the tower is constant. Or, the sum of the reboiler duty, plus the feed preheater duty, is constant. If the steam flow to the feed preheater is increased, then it follows that the reboiler duty will fall. How does this increase in feed preheat affect the flow of vapor through the trays and the fractionation efficiency of the trays?

The bottom part of the tower in Fig. 4.2, that is, the portion below the feed inlet, is called the *stripping section*. The upper part of the tower, that is, the portion above the feed inlet, is called the *absorption section*.

Since both the reflux flow and the overhead product flow are constant in this problem, it follows that the weight flow of vapor leaving the top tray is also constant, regardless of the feed preheater duty. Actually, this statement is approximately true for all the trays in the top or absorption part of the tower. Another way of saying this is that the heat input to the tower above the feed tray is a constant.

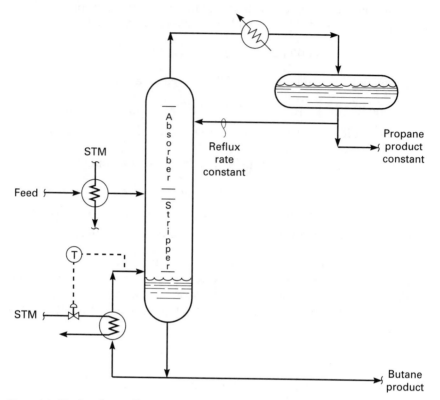

Figure 4.2   Feed preheat affects stripping efficiency.

But for the bottom stripping section trays, a reduction in reboiler duty will directly reduce the vapor flow from the reboiler to the bottom tray. This statement is approximately valid for all the trays in the stripping section of the tower.

As the flow of vapor through the absorption section trays is unaffected by feed preheat, the fractionation efficiency of the trays in the upper part of the tower will not change as feed preheat is increased. On the other hand, the reduced vapor flow through the stripping section may increase or decrease fractionation efficiency—but why?

Optimizing feed preheater duty

All types of valve and sieve trays are always suffering from lost tray efficiency, as a result of both flooding and dumping. Such trays always have some entrained droplets of liquid, lifted by the flowing vapors, to the trays above. This tends to blow butane up into the lighter propane product. Perforated trays always have some leakage of liquid through the tray deck to the trays below. This tends to drip propane down into the heavier butane product.

When we increase feed preheat and the reboiler duty is automatically reduced, dumping increases, but entrainment decreases. If the trays below the feed point were working poorly, because they were flooding, increasing feed preheat would improve their fractionation efficiency. If the trays below the feed point were working poorly because they were dumping, increasing feed preheat would reduce their fractionation efficiency. Figure 4.3 summarizes this effect. If, for this tower, we arbitrarily state that the percent of butane in the overhead propane product is constant, then the feed preheat duty, which minimizes the propane content in the butane bottoms product, represents the optimum preheater duty. We call this preheater duty the *incipient flood point*. The optimum feed preheater duty maximizes fractionation in a distillation tower at a fixed reflux rate.

Varying the enthalpy, or heat content, of the feed is an additional independent variable that an operator, or process design engineer, can use to optimize fractionation efficiency. An additional benefit of feed preheat is that a lower-level temperature heat source can be used. If valuable 100-psig steam is required for the reboiler, then low-value 20-psig steam might be adequate for the feed preheat exchanger.

Multicomponent systems

So far, all our examples have dealt with two-component systems. And many of our towers really just have two components. Also, we have assumed that the reflux rate is large compared to the overhead product rate. And many of our towers do run with a lot of reflux. But we can all

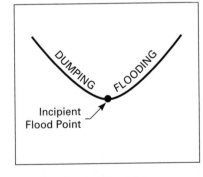

Figure 4.3 Optimizing feed preheat duty at a constant reflux rate.

think of distillation columns, where the top reflux rate is small compared to the overhead product, and the overhead product itself consists of a dozen widely different chemical compounds. Figure 4.4 represents such a column. It is called a *crude preflash tower*.

Notice that there is no reboiler in this flash tower. All the heat input comes from the partially vaporized crude. Both the temperature and the percent vaporization of the crude are fixed. Hence, the *external* heat input to the preflash tower is constant. The pounds of vapor flowing to the bottom tray must also be constant.

Now the overhead product of this tower is a mixture of a hundred different components, ranging from ethane, which has a molecular weight of 30, to decane, which has a molecular weight of 142. Also, while the overhead product rate is 60,000 lb/h, the top reflux rate is only 10,000 lb/h.

Consider the following. When the operator raises the top reflux rate, what happens to the weight flow of vapor going to the top tray? Recalling that the external heat input to this tower is constant, does the pounds per hour of vapor flowing to the top tray increase, remain the same, or decrease? The correct answer is *increase*. But why?

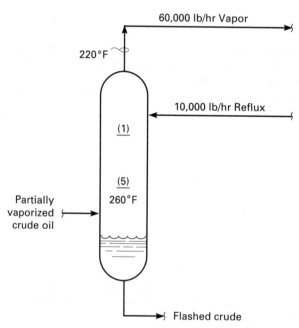

60,000 lb/hr Vapor

220°F

10,000 lb/hr Reflux

(1)

(5)
260°F

Partially
vaporized
crude oil

Flashed crude

Figure 4.4   Crude preflash tower.

Conversion of sensible heat to latent heat

When we raise the top reflux rate to our preflash tower, the tower-top temperature goes down. This is a sign that we are washing out from the upflowing vapors, more of the heavier or higher-molecular-weight, components in the overhead product. Of course, that is why we raised the reflux rate. So the reduction in tower-top temperature is good.

But what happened to the sensible-heat content (the heat represented by the temperature) of the vapors leaving the tower? As the vapor is cooler, the sensible-heat content decreased. But where did this heat go?

A small part of the heat was picked up by the extra liquid draining from the top tray. This extra liquid comes from the extra reflux. But the liquid flow through the tower is too small to carry away much heat. The main reason why the vapors leaving the top tray are cooler is vaporization; in other words, the sensible-heat content of the flowing vapors is converted to latent heat of vaporization.

But what is vaporizing? The reflux, of course. The sensible-heat content of the vapors, which is reduced when the reflux rate is increased, is converted to latent heat, as the vapors partially vaporize the incremental reflux flow.

As the reflux rate is raised, the weight flow of vapor through the top tray, and to a lesser extent through all the trays below (except for the bottom tray), increases. This increase in the weight flow of vapor occurs even though the external heat input to the preflash tower is constant. The weight flow of vapor to the bottom tray is presumed to be solely a function of the pounds of vapor in the feed.

Reduced molecular weight

A reduction in tower-top temperature of 20°F would increase the weight flow of vapor by roughly 10 percent. But the composition of the vapor would also change. The molecular weight of the vapor would drop by approximately 8 percent. The lower the molecular weight (MW) of a gas, the greater the volume, that a given weight of the gas occupies

$$\text{Gas volume} \approx \frac{\text{weight of gas}}{\text{MW of gas}}$$

In this equation, if the weight of gas goes up by 10 percent, and the molecular weight of the gas goes down by 8 percent, then the volume of gas goes up by 18 percent. The reduction in the tower-top temperature of 20°F does shrink the gas by about 2 percent, as a result of the temperature reduction, so that the net effect of raising the reflux rate is to increase the gas volume through the top tray by 16 percent (i.e., 18 percent minus 2 percent). This results in a substantial increase in the top-tray pressure drop, which can, and often does, cause the top tray to flood. This can happen, even though the external heat input and feed rate to the tower have never changed.

Internal reflux evaporation

The tray temperatures in our preflash tower, shown in Fig. 4.4, drop as the gas flows up the tower. Most of the reduced sensible-heat content of the flowing gas is converted to latent heat of evaporation of the downflowing reflux. This means that the liquid flow, or internal reflux rate, decreases as the liquid flows down the column. The greater the temperature drop per tray, the greater the evaporation of internal reflux. It is not unusual for 80 to 90 percent of the reflux to evaporate between the top and bottom trays in the absorption section of many towers. We say that the lower trays, in the absorption section of such a tower, are "drying out." The separation efficiency of trays operating with extremely low liquid flows over their weirs will be very low. This problem is commonly encountered for towers with low reflux ratios, and a multicomponent overhead product composition.

5

How Reboilers Work

Thermosyphon,
Gravity Feed,
and Forced

Four types of reboilers are discussed in this chapter:

- Once-through thermosyphon reboilers
- Circulating thermosyphon reboilers
- Forced-circulation reboilers
- Kettle or gravity-fed reboilers

There are dozens of other types of reboilers, but these four represent the majority of applications. Regardless of the type of reboiler used, the following statement is correct—almost as many towers flood because of reboiler problems as because of tray problems.

The theory of thermosyphon, or natural circulation, can be illustrated by the *airlift pump* shown in Fig. 5.1. This system is being used to recover gold bearing gravel, from the Magdalena River in Colombia, South America. Compressed air is forced to the bottom of the river through the air line. The air is injected into the bottom of the riser tube. The aerated water in the riser tube is less dense than the water in the river. This creates a pressure imbalance between points A and B. Since the pressure at point B is less than that at point A, water (as well as the gold and gravel) is sucked off the bottom of the river, and up into

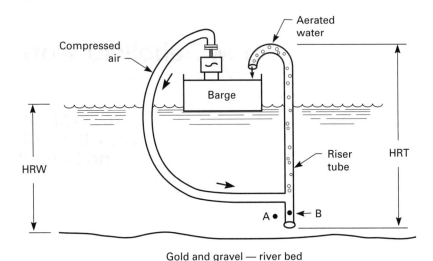

Figure 5.1 An airlift pump.

the riser tube. We can calculate the pressure difference between points
A and B as follows:

$$\frac{(\text{HRW})(\text{DRW}) - (\text{HRT})(\text{DRT})}{2.31} = \Delta P$$

where HRW = height of water above the bottom of the riser, ft
DRW = specific gravity of fluid in the riser; in this case 1.0
HRT = height of the aerated water in the riser tube, ft
DRT = specific gravity of aerated water in the riser tube (this
number can be obtained only by a trial-and-error calculation procedure)
ΔP = differential pressure between points A and B, psi

In a thermosyphon or natural-circulation reboiler, there is, of course,
no source of air. The aerated liquid is a froth or foam, produced by the
vaporization of the reboiler feed. Without a source of heat, there can be
no vaporization. And without vaporization, there will be no circulation.
So we can say that the source of energy that drives the circulation in a
thermosyphon reboiler is the heating medium to the reboiler.

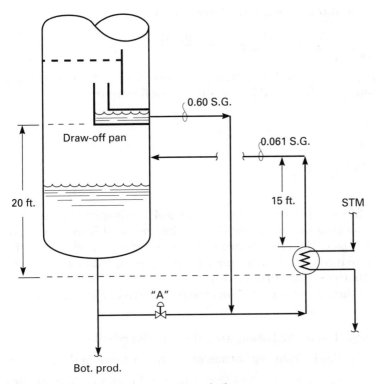

Figure 5.2 Once-through thermosyphon reboiler.

Thermosyphon Reboilers

Once-through thermosyphon reboilers

Figure 5.2 shows a once-through thermosyphon reboiler. The driving force to promote flow through this reboiler is the density difference between the reboiler feed line and the froth filled reboiler return line. For example:

- The specific gravity of the liquid in the reboiler feed line is 0.600.
- The height of liquid above the reboiler inlet is 20 ft.
- The mixed-phase specific gravity of the froth leaving the reboiler is 0.061.
- The height of the return line is 15 ft.
- Feet of water per psi = 2.31.

The differential pressure driving force is then

$$\frac{20 \text{ ft} \times 0.600 - 15 \text{ ft} \times 0.061}{2.31} = 4.7 \text{ psig}$$

What happens to this differential pressure of 4.7 psig? It is consumed in overcoming the frictional losses, due to the flow in the

- Reboiler
- Inlet line
- Outlet line
- Nozzles

If these frictional losses are less than the 4.7 psig given above, then the inlet line does not run liquid full. If the frictional losses are more than the 4.7 psig, the reboiler draw-off pan overflows, and flow to the reboiler is reduced until such time as the frictional losses drop to the available thermosyphon driving force.

The once-through thermosyphon reboiler, shown in Fig. 5.2, operates as follows:

- All the liquid from the bottom tray flows to the reboiler.
- None of the liquid from the bottom of the tower flows to the reboiler.
- All the bottoms product comes from the liquid portion of the reboiler effluent.
- None of the liquid from the bottom tray flows to the bottom of the tower.

This means that when the once-through thermosyphon reboiler is working correctly, the reboiler outlet temperature, and the tower bottoms temperature, are identical. If the tower-bottom temperature is cooler then the reboiler outlet temperature, then something has gone wrong with the thermosyphon circulation.

Loss of once-through thermosyphon circulation.   There are several common causes of loss of circulation. The common symptoms of this problem are

- Inability to achieve normal reboiler duty.
- Low reflux drum level, accompanied by low tower pressure, even at a low reflux rate.
- Bottoms product too light.

- Reboiler outlet temperature hotter than the tower-bottom temperature.

- Opening the steam or hot-oil inlet heat supply valve does not seem to get more heat into the tower.

The typical causes of this problem are

- Bottom tray in tower leaking, due to a low dry tray pressure drop.
- Bottom tray, seal pan, or draw-off pan damaged.
- Reboiler partially plugged.
- Reboiler feed line restricted.
- Reboiler design pressure drop excessive.
- Tower-bottom liquid level covering the reboiler vapor return nozzle.

If the loss of circulation is due to damage or leakage inside the tower, one can restore flow by opening the start-up line (valve A shown in Fig. 5.2), and raising the liquid level, but if the reboiler is fouled, this will not help.

Figure 5.3 shows a once-through thermosyphon reboiler with a *vertical baffle*. This looks quite a bit different from Fig. 5.2, but processwise, it is the same. Note that the reboiler return liquid goes only to the hot side of the tower bottoms. Putting the reboiler return liquid to the colder side of the tower bottoms represents poor design practice.

Circulating thermosyphon reboilers

This last statement requires some clarification. But to understand our explanation, an understanding of the important differences between a once-through thermosyphon reboiler, and a circulating thermosyphon reboiler, is critical. Figure 5.4 shows a circulating reboiler. In this reboiler

- The reboiler outlet temperature is always higher than the tower-bottom temperature.

- Some of the liquid from the reboiler outlet will always recirculate back into the reboiler feed.

- Some of the liquid from the bottom tray drops into the bottoms product.

- The tower-bottom product temperature and composition is the same as the temperature and composition of the feed to the reboiler.

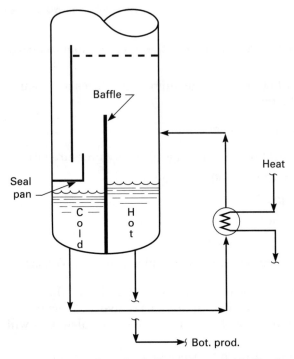

Figure 5.3 A once-through reboiler, with vertical baffle plate.

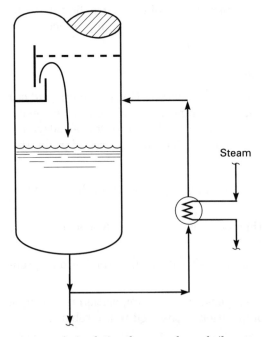

Figure 5.4 A circulating thermosyphon reboiler.

The liquid feed rate to the once-through thermosyphon reboiler is limited to the amount of liquid overflowing the bottom tray. The liquid feed rate to the circulating thermosyphon reboiler can be quite high—limited only by the available liquid head thermosyphon driving force. However, we should note that the liquid head thermosyphon driving force for a circulating thermosyphon reboiler is proportional to the height of the liquid level in the bottom of the tower above the reboiler inlet nozzle, whereas, with a once-through thermosyphon reboiler as described previously, the corresponding height is the elevation of the floor of the draw-off pan sump, above the reboiler inlet nozzle.

For a circulating thermosyphon reboiler, the rate of circulation can be increased by

■ Increasing the steam or hot-oil flow through the reboiler. This reduces the specific gravity or density of the froth or foam in the reboiler effluent line.

■ Increasing the tower bottoms liquid level. However, should this level reach the reboiler return nozzle, thermosyphon flow will be restricted, or even stop. Then, the reboiler heat duty will be reduced, and the tower pressure will drop.

Circulating vs. once-through thermosyphon reboilers

We said before that it was wrong to return the effluent, from a once-through reboiler, with a vertical baffle, to the cold side of the tower's bottom. Doing so would actually make the once-through thermosyphon reboiler work more like a circulating reboiler. But if this is bad, then the once-through reboiler, must be better than the circulating reboiler. But why?

1. The once-through reboiler functions as the bottom theoretical separation stage of the tower. The circulating reboiler does not, because a portion of its effluent backmixes to its feed inlet. This backmixing ruins the separation that can otherwise be achieved in reboilers.

2. Regardless of the type of reboiler used, the tower bottom product temperature has to be the same, so as to make product specifications. This is shown in Fig. 5.5. However, the reboiler outlet temperature must always be higher in the circulating reboiler than in the once-through reboiler. This means that it is more difficult to transfer heat in the former than in the latter.

3. Because the liquid from the bottom tray of a tower with a circulating thermosyphon reboiler is of a composition similar to that of the bottoms product, we can say that the circulating thermosyphon reboiler

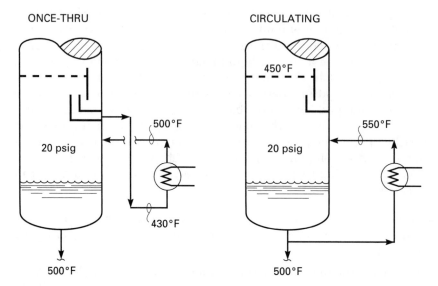

Figure 5.5 Once-through vs. circulating reboilers.

does not act as a theoretical separation stage. However, the liquid from the bottom tray of a tower with a once-through thermosyphon reboiler can be quite a bit lighter in composition (and hence cooler) than the bottoms product composition, and thus we say that the once-through thermosyphon reboiler does act as a theoretical separation stage. The cooler the liquid flow from the bottom tray of a tower, the less the vapor flow through that tray. This is because the hot vapor, flowing up through a tray, heats up the downflowing liquid. This means that there is a greater vapor flow through the bottom tray of a tower with a circulating thermosyphon reboiler, than there would be through the bottom tray of a tower with a once-through thermosyphon reboiler. Everything else being equal, then, the tower served by the circulating reboiler is going to flood before the tower served by the once-through reboiler.

Excessive thermosyphon circulation

In a once-through reboiler, the liquid flow coming out of the reboiler is limited to the bottoms product. In a circulating reboiler, the liquid flow coming out of the reboiler can be extremely high. If the reboiler return nozzle is located too close to the bottom tray of the tower, the greater volume of liquid leaving the nozzle can splash against the bottom tray. This alone can cause the entire column to flood. The best way to stop this flooding, is to lower the tower-bottom level.

Sometimes higher rates of thermosyphon circulation are good. They help prevent fouling and plugging of the reboiler, due to low velocity, dirt in the bottoms product, and especially high vaporization rates. If the percentage of vaporization in a once-through reboiler is above 60 percent, and dirt in the bottoms product is expected, then a circulating reboiler would be the better choice.

Forced-Circulation Reboilers

Figure 5.6 shows a once-through forced-circulation reboiler. Such a reboiler differs from a thermosyphon reboiler in that it has a pump to *force circulation,* rather than rely on natural or thermosyphon circulation. This seems rather wasteful—and it is.

The great advantage of forced circulation is that careful calculation of the pressure drop through the reboiler and associated piping is not critical. But, as we can see in Fig. 5.6, the operator now has two tower-bottom levels to control. Further, if the hot-side liquid level rises above

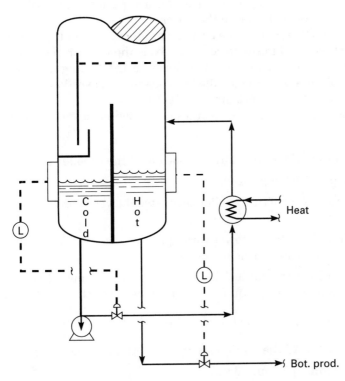

Figure 5.6   Forced-circulation once-through reboiler.

the reboiler return nozzle, the force of the vapor and liquid rushing back into the column will cause the trays to flood; but the reboiler heat input will not be affected.

Most often, forced circulation is used with fired reboilers. If flow is lost to such a reboiler, furnace tube damage is likely to result. Hopefully, this is less likely to occur with a forced-circulation reboiler. Also, the higher pressure drop of a furnace may force the designer to use a pump. Sometimes, we also see a forced-circulation reboiler system, if the reboiler heat is to be recovered from a number of dispersed heat sources that are far away from the tower, and hence a lot of pressure drop has to be overcome.

Kettle Reboilers

Reboilers are sometimes inserted into the bottom of a tower. These are called "stab-in" reboilers. It is not a terribly good idea, because it makes it more difficult to fix a leaking or fouled reboiler, without opening the tower itself. However, the "kettle" reboiler, shown in Fig. 5.7, has essentially the same process performance characteristics as the stab-in reboiler, but is entirely external to the tower.

Note that in a kettle reboiler, the bottoms product level control valve does not control the level in the tower; it controls the level on the product side of the reboiler only. The liquid level on the boiling or heat-exchanger side of the kettle is controlled by the internal overflow baffle. But what controls the tower-bottom liquid level?

To answer this, let us see how such a gravity-fed or kettle reboiler works:

- Liquid flows out of the tower, into the bottom of the reboiler's shell.
- The liquid is partially vaporized.
- The domed top section of the reboiler separates the vapor and the liquid.
- The vapor flows back to the tower through the riser line. This is the column's stripping vapor or heat source.
- The liquid overflows the baffle. The baffle is set high enough to keep the tubes submerged. This liquid is the bottoms product.

The level in the bottom of the tower is the sum of the following:

- The nozzle exit loss of the liquid leaving the bottom of the tower.
- The liquid feed-line pressure drop.

- The shell-side exchanger pressure drop, which includes the effect of the baffle height.
- The vapor-line riser pressure drop, including the vapor outlet nozzle loss.

Note that it is the elevation, or the static head pressure, in the tower that drives the kettle reboiler. That is why we call it a *gravity-fed reboiler*. Also, the pressure in the kettle will always be higher than the pressure in the tower. This means that an increase in the reboiler heat duty results in an increase of liquid level in the bottom of the tower.

Should the liquid level in the bottom of the tower rise to the reboiler vapor return nozzle, the tower will certainly flood, but the reboiler heat duty will continue. Unfortunately, reboiler shell-side fouling may also lead to tray flooding. This happens because the fouling can cause a pressure-drop buildup on the shell side of the reboiler.

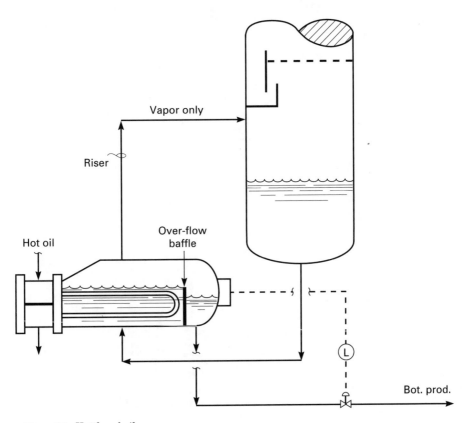

Figure 5.7 Kettle reboiler.

Remember, though, that the increased tower-bottom liquid level will not be reflected on the indicated bottom level seen in the control room, which is actually the level at the end of the kettle reboiler. This is a constant source of confusion to many operators, who have towers that flood, as a result of high liquid levels, yet their indicated liquid level remains normal.

6

How Instruments Work

Levels, Pressures, Flows, and Temperatures

In this age of advanced computer control, problems are still caused by the field instruments. The field instruments we discuss in this chapter are

- Level indicators
- Pressure indicators
- Flow indicators
- Temperature indicators

This chapter is particularly important when we consider that the data displayed in the control room are for operator control of the process. Data for engineering purposes should be obtained locally, at the instrument itself. Further, a large percent of control problems are actually process malfunction problems.

Level

Level indication

What is the difference between a *gauge glass* and a *level glass*? Simple! There is no such thing as a level glass. The liquid level shown in a

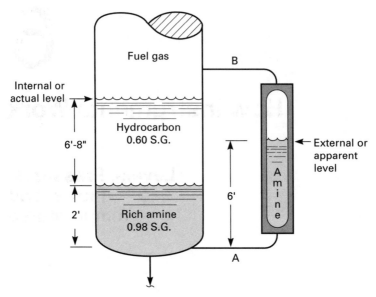

Figure 6.1 A gauge glass functions as a manometer.

gauge glass does not correspond to the level in a process vessel. Figure 6.1 is a good example. This is the bottom of an amine fuel-gas absorber. This tower is used to remove hydrogen sulfide from the fuel gas. At the bottom of the tower, there are three phases:

- Fuel gas: 0.01 specific gravity
- Hydrocarbon liquid: 0.60 specific gravity
- Rich amine: 0.98 specific gravity

Because of the location of the level taps of the gauge glass, only the amine is in the glass. The gauge glass simply measures the pressure difference between two points of the tower (points A and B in Fig. 6.1). That is, the gauge glass functions as a manometer, which measures the pressure difference, in terms of the specific gravity of the liquid in the gauge glass. Should the specific gravity of the liquid in the glass be the same as that of the liquid in the tower, then both the gauge-glass level and the tower level would be the same. But this is never so. The specific gravity of the liquid in the gauge glass is always greater than the specific gravity of the liquid in the tower. Hence, the apparent liquid level in the gauge glass is always somewhat lower than the actual liquid level in the tower.

This discrepancy between the *apparent level,* in the gauge glass, and the *actual level* (see Fig. 6.1), in the tower, also occurs in any other type of level-measuring device. This includes external float chambers, "kidneys," displacement chambers, and level-trols. The one exception to this is level-measuring devices using radiation techniques.

The three causes of the discrepancy between the external level and the internal level are

- Foam formation inside tower
- Ambient-heat loss from the external gauge glass or level-trol
- The liquid specific gravity in the glass, being greater than the specific gravity in the tower, as shown in Fig. 6.1.

Level discrepancies

Let's assume that the gauge glass shown in Fig. 6.1 holds 6 ft of amine. Since the bottom tap is in the amine phase and the top tap is in the gas phase, the liquid hydrocarbon is excluded from the gauge glass. To balance out the weight of the 6 ft of amine, the tower would have to have about 2 ft of amine and 6 ft 8 in of liquid hydrocarbon. That is, the tower liquid level would be about 8 ft 8 in or 2 ft 8 in higher than the gauge-glass level.

If you conclude from the above that we could use the gauge-glass level, to actually calculate the level inside the tower, you are quite wrong. To perform the preceding calculation, one would have to assume the *ratio* of *the phases.* But this is an assumption equivalent to assuming the answer. How, then, does one determine the actual liquid level in the tower, on the basis of the apparent liquid level in the gauge glass? The answer is that there is no answer. It cannot be done! And this statement applies to all other sorts of level-measuring instruments—with the exception of radiation devices.

Effects of temperature on level

The gauge glass will normally be somewhat colder than the process vessel as a result of ambient-heat losses (an exception to this would be a refrigerated process). For every 100°F decrease in the gauge-glass temperature or level-trol temperature, the specific gravity of the liquid in the glass increases by 5%. This rule of thumb is typical for hydrocarbons only. Aqueous (water-based) fluids are totally different.

For example, suppose the height of liquid in a gauge glass is 4 ft between the level taps. The glass temperature is 60°F. The tower temperature is 560°F. How much higher is the height of liquid in the tower than in the glass? (answer: 1 ft).

Explanation

$$500°F \times \frac{5}{100} \times \frac{1}{100°F} = \frac{25}{100}$$

- This means that the liquid in the gauge glass is 25% more dense than the liquid in the tower bottom.
- Assuming a linear relationship between density and volume, the level of liquid in the tower above the bottom tap of the gauge glass must be

$$\left(1 + \frac{25}{100}\right) \times 4 \text{ ft} = 5 \text{ ft}$$

- In other words, the liquid in the tower is 1 ft above the level shown in the glass.

Plugged taps

How do plugged level-sensing taps affect the apparent liquid level in a vessel? Let's assume that the vapor in the vessel could be fully condensed at the temperature in the gauge glass. If the bottom tap is closed, the level will go up, because the condensing vapors cannot drain out of the glass. If the top tap is closed, the level will go up, because the condensing vapors create an area of low pressure which draws the liquid up the glass through the bottom tap. Thus, if either the top or bottom taps plug, the result is a false high-level indication.

High liquid level

In our calculation above, we had 4 ft of liquid in the glass and 5 ft of liquid in the tower. But what happens if the distance between the two taps is 4 ft 6 in? I have drawn a picture of the observed result in Fig. 6.2. Liquid circulates through the glass, pouring through the top tap, and draining through the bottom tap. The apparent liquid level would then be somewhere between 4 ft 0 in and 4 ft 6 in; let's say 4 ft 2 in. The indicated liquid level on the control room chart would then be 92 percent (i.e., 4 ft 2 in÷4 ft 6 in). As the liquid level in the tower increases from 5 ft to 1000 ft, the indicated liquid level would remain at 92%.

Once the actual liquid level inside the tower bottom rises above the top-level tap, no further increase in level can be observed in the gauge glass.

The same sort of problem arises in a *level-trol,* which measures and transmits a process vessel liquid level to the control center. As shown in Fig. 6.3, the level-trol operates by means of two *pressure transduc-*

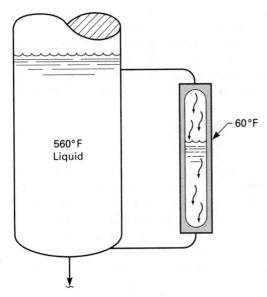

Figure 6.2   Liquid circulation through a gauge glass.

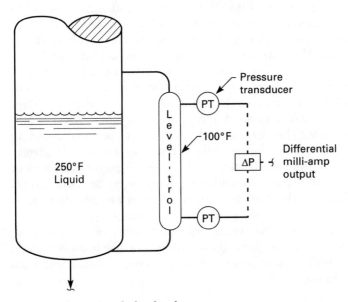

Figure 6.3   Operation of a level-trol.

ers, devices for converting a pressure signal into a small electric current. The difference between the two pressure transducers shown in Fig. 6.3 is called the *milliamp* (milliampere) *output.* This milliamp output is proportional to the pressure difference between the bottom and top taps in the level-trol. To convert the milliamp output signal from the level-trol into a level indication, the instrument technician must assume a specific gravity:

$$\text{Percent level} \sim \frac{\text{milliamp signal}}{\text{specific gravity}}$$

But which specific gravity should the instrument technician select? The specific gravity of the liquid in the level-trol, or the lower specific gravity of the liquid in the hotter process vessel? The technician should use the specific gravity in the process vessel, and ignore the specific gravity of the liquid in the level-trol. This can be especially confusing, if the operator then compares the apparently low liquid level in the gauge glass to the indicated higher liquid level on the control panel.

Foam Affects Levels

This procedure would work only if one knew the actual specific gravity of the fluid in the bottom of a distillation tower. But anyone who has ever poured out a glass of beer realizes that this is not possible. For one thing, the ratio of white froth to yellow beer is never known in advance. Also, the density of the froth itself is unknown, and is quite variable.

Figure 6.4 shows a distillation tower served by a circulating thermosyphon reboiler. To some unknown extent, some foam will always be found in the bottom of such vessels. Not sometimes, but always. Why?

The purpose of a tray is to mix vapor and liquid. This produces aerated liquid—or foam. The purpose of a reboiler is to produce vapor. In a circulating reboiler, the reboiler effluent flows up the riser as a froth. Of course, the flow from the bottom of the tower is going to be a clear liquid. Foam cannot be pumped. But there will always be some ratio of foam to clear liquid in the bottom of the tower, and we have no method of determining this ratio, or even the density of the foam.

Well, if we do not know the average specific gravity of the foamy liquid in the bottom of a tower, how, then, can we find the level of foam in the tower? The unfortunate answer is that, short of using radiation techniques, we cannot.

Split liquid levels

The two gauge glasses shown in Fig. 6.4 both show a liquid level. Many of you may have observed this on a process vessel. We certainly cannot

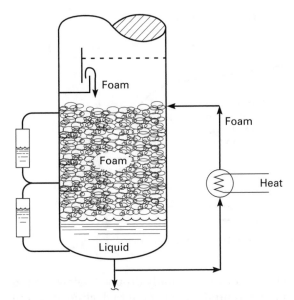

Figure 6.4 Split liquid-level indication caused by foam.

have layers of liquid-vapor-liquid-vapor in the vessel. Rather, these split liquid levels are a positive indication of foam or froth in the bottom of the tower.

If the foam is spanning both taps on a gauge glass, then the height of the liquid in the glass is a measure of the specific gravity, or density of the foam, in terms of the specific gravity of the liquid in the glass. If the foam is above the top tap of both the gauge glasses in Fig. 6.4, then there will be a level in both glasses. The upper gauge glass will show a lower level because the light foam in the tower floats on the top of the heavier foam. Note that these split liquid levels, so often seen in a process vessel, tell us nothing about the real liquid level in the vessel. They are a sign of foam.

Figure 6.5 is a plot of the liquid level in a crude preflash drum versus time. We were steadily withdrawing 10 percent more flashed crude from the bottoms pump, than the inlet crude feed rate. The rate of decline of the liquid level noted in the control center was only about 25 percent of our calculated rate. Suddenly, when the apparent level in the control room had reached 40 percent, the level indication started to decline much more rapidly. Why?

This extreme nonlinear response of a level, to a step change in a flow rate, is quite common. Before the sudden decline in the liquid level, foam had filled the drum above the top-level tap. The initial slow decline in the apparent level was due to a dense foam dropping between

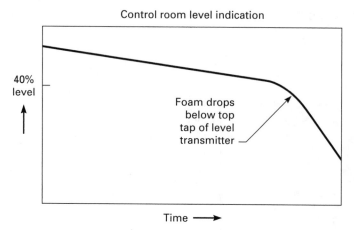

Figure 6.5 Foam creates a nonlinear response in level indication.

the level taps. Only when the foam level actually dropped below the top tap of the drum did the indicated liquid level begin to decline at a rate representing the actual decline in the level. Thus, we can see that this common, nonlinear response, is not due to instrument malfunctions, but is a sure sign of foam or froth.

Radiation level detection

The only way around the sort of problems discussed above is to use neutrons or x-rays, to measure the density in a vessel. In a modern petroleum refinery, perhaps 5 percent of levels are monitored with radiation. It is both safe and effective.

The neutron backscatter technique is best performed with hydrogen-containing products. Both the source of a slow neutron and the receiver are located in the same box. The slow neutrons bounce off of protons (hydrogen ions) and are reflected back. The rate at which these neutrons are reflected back is measured, and corresponds to the hydrocarbon density in the vessel. This measurement is not affected by steel components inside—or outside—the vessel.

X-ray level detection works with a source of radiation and a receiver, located on either side of the vessel. As the percent absorption of the radiation increases, the receiver sees fewer x-rays, and a higher density is implied. The x-rays are absorbed by steel components, such as ladders and manways, which can sometimes be confusing.

Either method discriminates nicely between clear liquid, foam, or vapor. Such a level controller can be calibrated to hold a foam level, or a liquid level. Of course, this sort of level detection is far more expensive than conventional techniques.

Pressure

Pressure indicators

The chief engineer of a West Texas process plant had decided to replace the main condenser. Colder weather always coincided with a vastly improved vacuum, in their vacuum tower. It seemed as if colder air to the condenser really helped. So the engineering concluded, that a bigger condenser would also help during warm weather.

Wrong! The chief engineer failed to realize that the vacuum pressure indicator was not equipped with a *barometric pressure compensator.* An ordinary vacuum pressure indicator or pressure gauge reads the pressure difference between the vacuum system and atmospheric pressure. When ambient temperatures drop, the barometer rises or ambient pressure goes up. An ordinary vacuum pressure gauge or indicator would then read an improved vacuum. But in reality, the vacuum has not changed.

The opposite problem would occur in Denver—the Mile-High City. At sea level, full vacuum is 30 in of mercury (or 30 in Hg). But in Denver, full vacuum is about 26 in Hg. An ordinary vacuum pressure gauge reads zero inches of mercury (0 in Hg) in Denver and in New Orleans, because although these cities are at different altitudes, the gauge compares system pressure only with ambient pressure. But a vacuum pressure gauge reading 25 in Hg in New Orleans would correspond to a poor vacuum of 5 in Hg absolute pressure (30 in Hg−25 in Hg). A vacuum pressure gauge reading 25 in Hg in Denver would correspond to an excellent vacuum of 1 in Hg absolute pressure (26 in Hg−25 in Hg).

All these complications can be avoided when making field measurements by using the vacuum manometer shown in Fig. 6.6. The difference between the two mercury levels is the absolutely correct, inches of mercury absolute pressure, or millimeters of mercury (mm Hg).

Pressure transducers

Disassemble a pressure transducer, and you will see a small plastic diaphragm. A change in pressure distorts this diaphragm, and generates a small electrical signal. The signal must be quite tiny, because placing your hand on the transducer can alter its reading. A modern digital pressure gauge uses a pressure transducer. This type of gauge, if zeroed at sea level in New Orleans, will read 4 in Hg in Denver. Most pressure signals transmitted from the field into the control center are generated from pressure transducers. Differential pressure indicators simply take the differential readings from two transducers and generate a milliamp output signal.

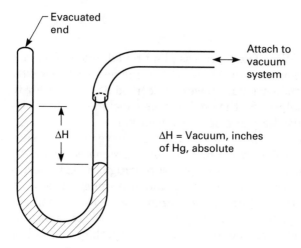

Figure 6.6   A mercury absolute-pressure manometer.

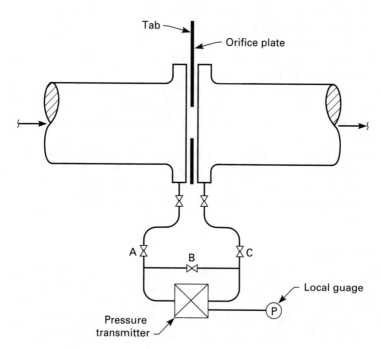

Figure 6.7   Flowmeter assembly.

Pressure-point location

Locating a pressure tap in an area of high velocity is likely to produce a lower pressure indication than the real flowing pressure. Using a purge gas to keep a pressure tap from plugging often can cause a high pressure reading, if too much purge gas or steam is used. A pressure tap located below a liquid level will read too high; pressures should be measured in the vapor phase.

Flow

Flow indication

The standard method of measuring flows in a process plant is by use of the orifice plate and orifice flanges, shown in Fig. 6.7. Actually, we rarely measure flows directly. More commonly, we measure the pressure drop across an orifice plate. This pressure drop is due to the increase in kinetic energy of a fluid as it accelerates through the small hole in the orifice plate. The energy to provide the increased velocity comes from the pressure of the flowing fluid, in accordance with the following:

$$\Delta P = K \frac{D_f}{62.3} V^2$$

where ΔP = measured pressure drop through the orifice plate, in inches of water (multiply the measured pressure drop in psi by 27.7 to obtain the inches of water ΔP)

V = velocity of the fluid through the orifice plate, ft/s
D_f = density of the fluid, whether vapor or liquid, lb/ft^3
K = an orifice coefficient

You should look up the orifice coefficient K in your Cameron or Crane handbook—but it is typically a number like 0.6 to 0.8.

Checking flows in the field

The competent engineer does not assume a flow indication shown on the control panel is correct, and proceeds as follows:

- Referring to Fig. 6.7, place an easy-to-read pressure gauge in the position shown. I like to use a digital gauge.
- By opening both valves A and B, with C closed, you will now be reading the upstream pressure.
- By opening valve C, with A and B closed, you will read the pressure downstream of the orifice plate.

- The difference between the two readings is ΔP, in the preceding equation. Now solve for V.

- Look at the tab sticking out of the orifice flanges (see Fig. 6.7). If the orifice plate is installed in the correct direction, there will be a number stamped onto the tab, toward the flow. This is the orifice plate hole diameter; for example, if you see 0.374" stamped on the tab, the orifice hole diameter should be 0.374 in.

- Using the hole diameter, calculate the volume of fluid.

You may notice, when you measure ΔP, that it is a small value, difficult to measure accurately. This means that the orifice plate hole is oversized, and that the accuracy of the recorded flow on the control panel is also poor. Or, the measured ΔP is quite high. This means that a lot of pressure is being wasted, and the orifice plate hole is undersized and restricting flow. Furthermore, the recorded flow on the control panel may be off scale.

The reason the orifice flanges are kept close to the orifice plate is that when the liquid velocity decreases, downstream of the orifice plate, the pressure of the liquid goes partly back up. Figure 6.8 illustrates this point. It is called *pressure recovery*. Whenever the velocity of a flowing fluid (vapor or liquid) decreases, its pressure goes partly back up. An extreme example of this is *water hammer*. The reason the pressure at the end of the pipe in Fig. 6.8 is lower than at the inlet to the pipe is due to frictional losses.

The orifice coefficient K takes into account both frictional pressure losses, and conversion of pressure to velocity. The frictional losses represent an *irreversible process*. The conversion of pressure to velocity represent a *reversible process*.

Other flow-measuring methods

A better way of measuring flows, than the ordinary orifice plate method, is by inducing vortex shedding across a tube in the flowing liq-

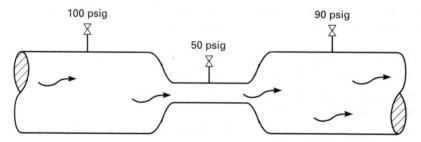

Figure 6.8 Pressure recovery for water flow in pipe.

uid, and then measuring the velocity of the vortices. This is a nice method, as there are no orifice taps to plug. Then there are Doppler meters, which measure the velocity of a fluid, based on how the speed of sound is affected by the flow in a pipe. More commonly, we have rotometers, which measure how far a ball or float is lifted, in a vertical tube by the velocity of the liquid. But regardless of their relative merits, perhaps simply for historical reasons, the vast majority of flows, in most process plants, are measured with the orifice plate flowmeter, shown in Fig. 6.7.

Glycol-filled instrument lines

Many of us have seen the following tag attached to level-sensing lines, or to a flow transmitter: "Do not drain—glycol-filled." This means the instrument mechanic has filled the lines with glycol, mainly for winter freeze protection. Many process streams contain water, which can settle out at low points and, in effect, plug the impulse lines to flow- or level-sensing ΔP transmitters, when water freezes. Note that there is not a lot of difference between measuring a flow and a level; they both are typically measured by using a differential pressure signal.

Naturally, just like level indicators, the flow orifice taps can plug. If the upstream tap plugs, the flow will read low or zero. It is best to blow the tap back with glycol, but that is not always practical. If you blow the taps out with the pressure of the process stream, you do not need to refill the impulse lines with glycol to get a correct flow reading. But the lines have to be totally refilled with the same fluid. If you are measuring the flow of a single-phase liquid, just open valves A, B, and C (shown in Fig. 6.7) for a few minutes. If you are working with vapor at its dew point or wet gas, there is a problem. If the flow transmitter is located below the orifice flanges, you will have to wait until the impulse lines refill with liquid. Open valve B, and close valves A and C. Now wait until the flowmeter indication stops changing. It ought to go back to zero, if the lines are refilled.

Zeroing out a flowmeter

The indicated flow of acetic acid is 9000 liters per day. The instrument technician checks the flowmeter to see if it has drifted, by opening valve B, with A and C closed (see Fig. 6.7). It should go back to zero—but a reading of 2000 liters per day is noted. The full range on the flowmeter is 10,000 liters per day. What is the real flow rate of the acetic acid? The answer is not 2000 liters. Why? Because flow varies with the square root of the orifice plate pressure drop. To calculate the correct acetic acid flow:

- $9000^2 - 2000^2 = 77,000,000$
- $(77,000,000)^{1/2} = 8780$ liters per day

The lesson is that near the top end of its range, the indicated flow is likely to be accurate, even if the meter is not well zeroed, or the measured ΔP is not too accurate. On the other hand, flowmeters using orifice plates cannot be very accurate at the low end of their range, regardless of how carefully we have zeroed them. Digitally displayed flows also follow this rule.

Temperature

Temperature indication

Figure 6.9 shows an ordinary thermowell-and-thermocouple assembly. The thermocouple junction consists of two wires of different metals. When this junction of the wires is heated, a small electric current, proportional to the junction temperature, is produced. Different metal wires make up the three most common junctions: *J, H,* and *K.* It is not uncommon for a thermocouple, regardless of the type of junction, to generate too low a temperature signal.

If the exterior of the thermowell becomes fouled, the indicated temperature, generated by the thermocouple, will drop. The problem is that the external cap of the thermowell assembly radiates a small amount of heat to the atmosphere. Normally, this has a negligible effect on the indicated temperature. However, when the process temperature is 600 to 800°F, the thermowell is in a vapor phase, and the thermowell becomes coated with coke, I have seen the indicated tem-

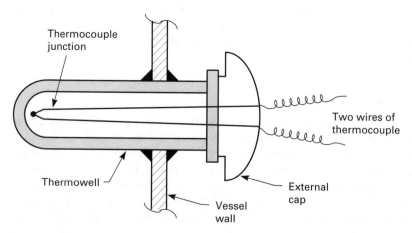

Figure 6.9 Thermocouple assembly.

perature drop by 40°F below its true value. To verify that fouling of a thermowell is a problem, place a piece of loose insulation over the exterior thermowell assembly. If the indicated temperature rises by 5 or 10°F, then fouling on the outside of the thermowell is proved.

Short thermowells

For a thermocouple to read correctly, it should be fully inserted in a thermowell and the thermowell itself should extend several inches into the process liquid. If the process stream is a vapor, which has poorer heat-transfer properties than do liquids, the thermowell, especially if the external insulation is poor, should extend more than 6 in into the process flow. To check the length of the thermowell, unscrew the thermocouple assembly and pull it out, then simply measure the length of the thermocouple. This is also a good opportunity to verify the control-room reading with a portable temperature probe, or a glass thermometer, inserted in the thermowell. About 5 percent of the TI points used are not located where the unit piping and instrumentation diagrams (P&IDs) show them to be. Pulling the thermocouple from a point in the process sometimes causes an unexpected drop to ambient temperature, at an entirely unexpected TI location.

Ram's-horn level indication

A too short thermowell in the side of a vessel invariably exhibits an increase in temperature, when the liquid level rises to submerge the thermowell. Figure 6.10 illustrates a common method of exploiting this

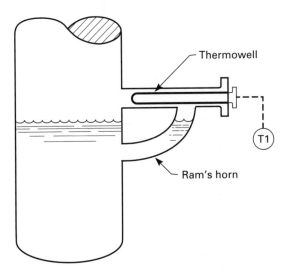

Figure 6.10   Using temperature to sense a liquid level.

phenomenon. This is the ram's-horn interface-level indicator. The thermowell extends to the vessel wall only, and is poorly insulated. The curved pipe below the thermowell permits liquid to drain out of, and through, the pipe enclosing the thermowell. Usually, three or four such ram's horns are vertically set, 18 in apart. A sudden temperature increase at a ram's horn is a foolproof method of detecting a rise in a liquid level. Especially in fouling or plugging service, I have seen this simple, archaic method of level indication succeed when all else fails.

If two temperature readings from the same point in the process disagree, the chances are that a temperature indication is more accurate than a temperature-control signal. The temperature signal used for control has usually been converted from its direct milliamp output to operate a control station. The temperature indication is generated right from the junction of the thermocouple, and hence there is less chance for error.

7

Packed Towers:
Better than Trays?

Packed-Bed Vapor and
Liquid Distribution

The very first continuous distillation column was the patent still used to produce Scotch whiskey in the 1830s. It had 12 bubble-cap trays with weirs, downcomers, tray decks, and bubble caps with internal risers. Current trayed towers are quite similar. As most distillation towers have always been trayed rather than packed, one would have to conclude that trayed towers must have some sort of inherent advantage over packed towers. And this is indeed true, in a practical sense; even though, in theory, a packed tower has greater capacity and superior separation efficiency than a trayed column.

How Packed Towers Work

The original packed towers used *Raschig rings,* hollow ceramic cylinders, typically 1 in outer diameter (OD), ¾ in inner diameter (ID), and 1 in long. A state-of-the-art packing is called *structured packing.* This material is made of thin sheets of crimped or corrugated metal, with small holes, fastened together with thin bars. The metal selected for the sheets is typically 316 stainless steel.

The two features that should be maximized in packed beds are:

- *Open area—the average percentage of the cross-sectional area of the tower not blocked by the packing, and hence available for the flow of vapor and liquid.*

- *Wetted surface area*—the number of square feet of packing surface area available for vapor-liquid contacting, per cubic foot of tower volume.

The greater the open area of a packing, the greater the capacity of a tower. The greater the wetted surface area of a packing, the greater the separation efficiency of the tower. For example, a packing consisting of empty space, would have lots of capacity, but awful separation efficiency. A packing consisting of a fine sand would have great separation efficiency, but very low capacity. So, the selection of packing for a column is a compromise between maximizing open area and maximizing the wetted surface area.

Structured packing has about 50 percent more open area than Raschig rings and two or three times their wetted surface area. Hence, structured packing has largely replaced packing in the form of rings, in many packed towers.

In any type of packed tower, the liquid or internal reflux drips through the packing, and forms a thin film of liquid on the surface of the packing. Vapor percolates up through the packing, and exchanges heat and molecules with the thin film of liquid on the surface of the packing.

In a trayed tower, vapor-liquid contact occurs, only on the 5 or 6 in above the tray deck, and the majority of the tower's volume is not used to exchange heat or mass between vapor and liquid. In a packed tower, the entire packed volume is used for this vapor-liquid contacting.

In a trayed tower, the area used for the downcomer which feeds liquid to a tray, and the area used for draining liquid from a tray, are unavailable for vapor flow. In a packed tower, the entire cross-sectional area of the tower is available for vapor flow.

This certainly makes it seem as if packing is vastly superior to trays, and even if this was not true in the past, with the advent of structured packings in the 1980s, it ought to be true now. But as a process design engineer, whenever possible, I still specify trays, rather than packing. Why?

Liquid distribution

Each tray in a tower is inherently a vapor-liquid *redistributor*. The outlet weir, or more exactly the crest height of the liquid overflowing the weir forces the liquid to flow evenly across the tray. Even if the weir

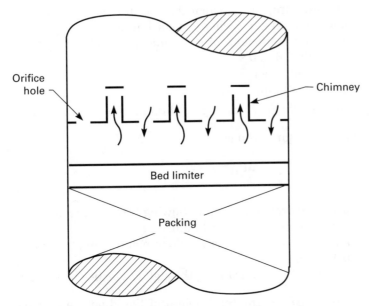

Figure 7.1 Orifice plate liquid distributor.

height is uneven on a tray, and liquid flow is distorted on that particular tray, the liquid will be properly redistributed, on the tray below.

Unfortunately, packing does not redistribute liquid, or internal reflux. Unless the initial reflux distribution is good, the liquid flow distribution through the entire packed bed will be poor. Figure 7.1 shows a common orifice plate liquid distributor. Vapor flows up through the large chimneys, and liquid drains through the smaller distribution holes in the tray deck.

To calculate the height of liquid on this tray, we have to add two factors:

- The pressure drop of the vapor flowing through the chimneys. The liquid on the tray has to develop enough liquid head to flow against the higher pressure below the tray.

- The orifice hole pressure drop. This is the liquid head, which has to be converted to velocity as the liquid flows through the orifice holes on the tray deck.

The pressure drop of the vapor flowing through the chimneys is

$$\Delta P_{\mathrm{V}} = K_c \frac{D_v}{D_l} (V_g)^2$$

where ΔP_V = pressure drop of the vapor flowing through the chimneys and hats, in inches of clear liquid

$\quad K_c$ = an orifice coefficient for a chimney plus a chimney hat, typically 0.6 to 0.9

$\quad D_v$ = density of vapor

$\quad D_l$ = density of liquid

$\quad V_g$ = velocity of vapor passing up through the chimneys, ft/s

The orifice hole pressure drop is

$$\Delta P_L = K_0 \, V_l^2$$

where ΔP_L = pressure drop of the liquid flowing through the orifice holes, in inches of clear liquid.

$\quad K_0$ = an orifice coefficient for a hole in metal plate, typically 0.4 to 0.6

$\quad V_l$ = velocity of liquid draining through the orifice holes, ft/s

The sum of $\Delta P_V + \Delta P_L$ is the total height of liquid on the chimney tray. Let us assume that this total height is 6 in of liquid:

- $\Delta P_V = 5.1$ in
- $\Delta P_L = 0.9$ in

Let's further assume that the orifice plate distributor is 1 in *out-of-level*. This could easily happen in a 14-ft 0-in-ID tower. Figure 7.2 shows the results. The flow of internal reflux or liquid through the higher portion of the tray deck falls to zero. Worse yet, vapor starts to

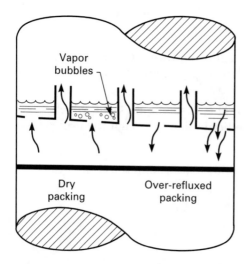

Figure 7.2 Orifice plate distributor not level.

bubble up through any orifice hole that is not submerged by at least 5.1 in of *clear liquid*. But, of course, once vapor bubbling begins, we do not have clear liquid in this region of the tray deck.

Meanwhile, all the liquid flow must drain through that portion of the tray that is lower. The net result, as can be seen in Fig. 7.2, is dry packing in one portion of the tower, and overrefluxing in the adjacent portion. This is called *vapor-liquid channeling,* and this is the root cause of poor fractionation efficiency in any tower.

One way out of this problem, is to increase ΔP_L, the pressure drop of the liquid flowing through the orifice holes. This could be done, by increasing the orifice hole liquid velocity. We could drill fewer orifice holes. Unfortunately, this would decrease the number of *drip points* per square foot of tower area (6 to 10 is a good target). This would reduce vapor-liquid contacting efficiency. Or, we could have smaller orifice holes. But too small a hole would probably plug with corrosion products.

There are ways around these problems. Figure 7.3 is a simplified sketch of a "multipan, liquid distributor" using "raised, slotted, guided tubes" employing an "initial distribution header." It is beyond the scope of this text to describe how this works. But I will say that such liquid distributors are:

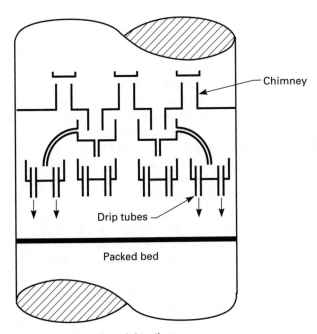

Figure 7.3 Modern liquid distributor.

- Expensive to build
- Tricky to design—especially for small liquid flows
- Difficult to install correctly
- Occupy quite a few feet of tower vertical height

Tradeoff between efficiency and capacity. In a trayed tower, we can increase tower capacity by reducing the number of trays, and increasing the *tray spacing*. But reducing the number of trays in turn reduces separation efficiency. In a packed tower, we can increase tower capacity by using larger packing and increasing open area (or a bigger crimp size for structured packing). But the reduction in the wetted area of the packing, will reduce separation efficiency. Installing a properly designed, efficient liquid distributor, such as the one shown in Fig. 7.3, will definitely enhance separation efficiency, but at the expense of the vertical height of the tower, which could otherwise be devoted to a greater height of packing.

Vapor distribution

Figure 7.4 shows a modern, narrow-trough, liquid collector-vapor distributor chimney tray. While the initial vapor distribution through a packed bed is not quite so critical or difficult as the liquid distribution, it is still important.

A properly designed valve or sieve tray will act as a vapor redistributor. Thus, poor initial vapor distribution will only lessen the efficiency of the bottom tray. But if a packed-bed vapor distributor does not work properly, vapor channeling will be promoted through the entire bed.

The narrow-trough vapor distributor shown in Fig. 7.4 is intended to disperse the vapor evenly across the bottom of the packed bed. The width of the chimney does not exceed 6 in. The older-style chimney trays, which may have had a few large round or square chimneys, reduced the separation efficiency of the packing. To work properly, the vapor distributor has to have a reasonable pressure drop, in comparison to the pressure drop of the packed bed. For example, if the expected pressure drop of a 12-ft packed bed is 10 in of liquid, the pressure drop of the vapor distributor ought to be about 3 to 4 in of liquid.

In practice, the most common reason why vapor flow is maldistributed as it escapes from a well-designed chimney tray is distortion of the hats. I believe this is due mostly to workers stepping on the hats during unit turnarounds. The hat support brackets are designed to support a 20-lb hat, not a 200-lb welder.

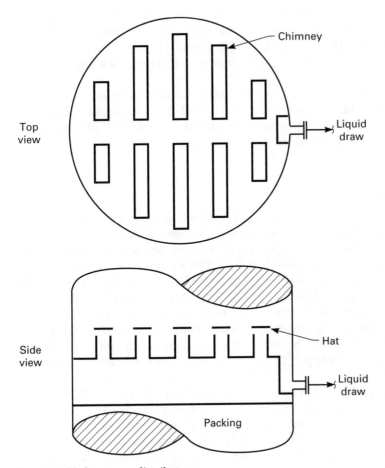

Top view

Side view

Figure 7.4 Modern vapor distributor.

Maintaining Functional and Structural Efficiency in Packed Towers

Pressure drop

A packed tower can successfully fractionate with a very small pressure drop, as compared to a tray. For a modern trayed tower, to produce one single theoretical tray worth of separation (that's like a single, 100 percent efficient tray), a pressure drop of about 6 in of liquid is needed. A bed of structured packing can do the same job, with one inch of liquid pressure drop, even when allowing for the vapor distributor. In low-pressure fractionators, especially vacuum towers used to make lubricating oils and waxes, this can be of critical importance.

Packed beds also seem to have a better turndown capability than valve or sieve trays, at low vapor flows. On the other hand, many packed fractionators seem quite intolerant of reduced liquid or reflux flow rates. This is typically a sign of an improperly designed distributor in the packed fractionator.

The problem we have just discussed—poor fractionation efficiency due to inadequate vapor and liquid initial distribution—is rather similar to tray deck dumping in trayed fractionators. And, just like trays, packed towers are also subject to flooding.

Flooding in packed towers

To understand what is meant by flooding in a bed of packing, one ought to first understand what is meant by the idea of *holdup.*

Let's imagine we are operating the air-water scrubber in the Unit-Ops Lab at Cooper Union (Fig. 7.5). The water is circulating from the bottom to the top of the tower. The water level in the bottom is 2 ft. The height of the packed bed is 20 ft. At 3:30 P.M., Professor Liebskind shuts off our air compressor and the water circulation pump. As the water, which had been held up on the packing by the flowing air, drains down, the water level in the bottom of the tower rises from 2 to 5 ft. We say, then, that the holdup of the packing was 15 percent (5 ft − 2 ft ÷ 20 ft = 15 percent).

If the liquid holdup is too low, fractionation efficiency will be bad. We say that the *height equivalent to a theoretical plate* (HETP) will be high. If the liquid holdup is too high, fractionation efficiency will also be poor. We again say that the HETP will be high. This idea is expressed in Fig. 7.6. When the holdup rises above the point that corresponds to the minimum HETP, we can say that the packing is beginning to flood. The minimum HETP point on Fig. 7.6 corresponds to the point of incipient flood, discussed in Chap. 1.

For structured-type packing, a liquid holdup of 4 to 5 percent corresponds to this optimum packing fractionation efficiency. For 1-in Raschig rings, this optimum holdup would be roughly 10 to 12 percent.

There is a close relationship between percent liquid holdup and pressure drop, as expressed in inches of liquid per foot of packing. If the pressure drop per foot of packing is 1.2 in, then the liquid holdup is roughly 10 percent (1.2 in ÷ 12 in = 10 percent). Unlike the perforated tray decks discussed in Chap. 1, there is no simple equation to predict the optimum pressure drop that would correspond to the minimum HETP for different types of packing. There is just too much diversity in both size and shape to generalize. I can, however, recall a delayed coking fractionator in Los Angeles with a 7-ft bed of structured packing.

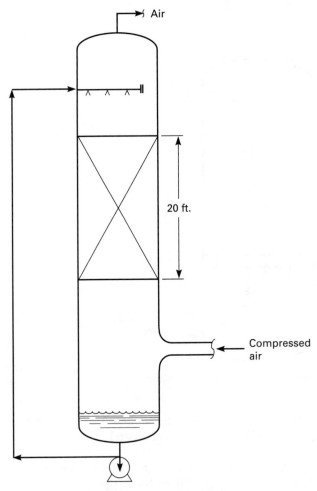

Figure 7.5 Measuring holdup in a packed bed.

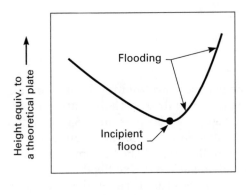

Figure 7.6 Holdup predicts flooding.

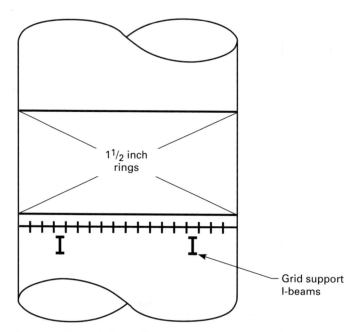

Figure 7.7   A restrictive packing support will cause flooding.

When the pressure drop per foot of packing reached 4 in (due to coke formation), drainage of liquid through the bed ceased.

Packed-bed supports.   A simple support for a bed of 1½-in pall rings might consist of a bar grid, with ¼-in-thick bars, set on 1 inch center. The grid itself (see Fig. 7.7) would be supported by two 6-in-wide I-beams. The open area of each component is

- 1½-in pall rings: 80 percent
- Grid support: 75 percent
- I-beam support: 90 percent

Considering that the rings, the grid support, and the I-beam are all in close physical contact, what is the open area at their mutual interface? (Answer: 80 percent × 75 percent × 90 percent = 54 percent.) This means that there will be a restriction to vapor and liquid flow at the grid support, which will promote flooding at the bottom of the bed. Naturally, this would cause the entire packed bed to flood.

To avoid such flooding, the designer uses the corrugated bed support shown in Fig. 7.8. This grid will have over 100 percent open area. Crushed packing laying on even this excellent type of bed support can, however, cause flooding.

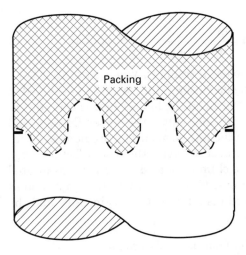

Figure 7.8   Superior bed support.

Packing holddowns

During routine operations, the weight of a bed of packing is sufficient
to keep it in place. But if there is a pressure surge during start-up, the
packed bed can be forcefully pushed against the underside of the liquid
distributor. If this liquid distributor is a spray header, as illustrated in
Fig. 7.9, the spray bar arms can be bent. To prevent this, a packing top
holddown grid is required. This holddown grid must be firmly attached
to the vessel wall. For example, in a 12-ft-ID tower, the holddown grid
is designed to resist an uplift force due to a pressure surge of 0.5 psi.

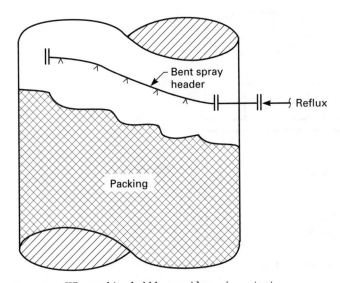

Figure 7.9   Why packing holddown grids are important.

What is the total uplift force?

$$\text{Answer: } 0.5 \text{ psi} \times 144 \, \frac{\text{in}^2}{\text{ft}^2} \times \frac{3.14}{4} \times 12^2 = 8000 \text{ lb}$$

Figure 7.10 illustrates a problem introduced when a packing hold-down was installed at the wrong elevation. This tower was designed to purify acetic acid, by distilling overhead the lighter formic acid. But at even 50 percent of the design reboiler duty, the tower flooded, and acetic acid was carried over into the formic acid. The problem was that the perforated plate holddown, used to keep the 1½-in Teflon rings from floating up through the feed chimney tray distributor, was located immediately below these chimneys. What, then, was the open area at the interface of the chimney feed distributor, the 1½-in rings and the holddown plate?

- Open area of chimney feed distributor = 40 percent
- Open area of rings = 70 percent
- Open area of holddown plate = 30 percent (Answer: 40 percent × 70 percent × 30 percent = 8.4 percent.)

As a result, the tower flooded above the holddown plate. To fix this problem, the packing holddown was dropped 15 in below the chimney, orifice plate, distributor. As a result, the tower fractionated properly.

Incidentally, this incident was a multi-million-dollar mistake. It was not, however, a design error. The installation crew had simply found a

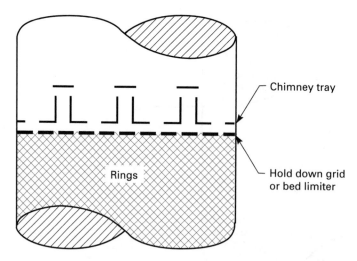

Figure 7.10 Flooding caused by mislocation of holddown grid.

"cheaper" location to install the holddown grid, by using the same support as the chimney feed distributor.

Crushed packing

One of the great disadvantages of packing is the inability to properly inspect the installation. I well remember one coal gasification project in Chicago, where the off-gas scrubber flooded because of a plastic bag left inside a packed bed of pall rings. On another tower, flooding of a bed-structured packing occurred at 50 percent of design. Only when the packing was removed for inspection was the problem revealed. Workers had previously stepped on and crushed intermediate layers of packing. Then, probably without realizing that they had reduced the open area of the structured packing by 50 percent, they set additional layers of structured packing on top of the damaged layers.

The situation on a trayed tower is far more controllable. Even after the tray assembly is complete, the process engineer, using only a crescent wrench, can open a few tray deck manways and double-check the tower internals. But to pull 6 ft of packing out of a tower is a major undertaking. Also, trays can be stepped on without damage. Beam supports and 40-lb airguns can be dragged across a sieve tray with impunity. Packings—rings or structured—require special handling during installation or turnarounds. To be frank, all packing installations that I have seen have been damaged to some extent.

To summarize, the chief disadvantages of packed towers, as compared to trays, are

- Difficulty of achieving proper initial liquid distribution, especially at small liquid flow rates
- The need for a proper vapor distributor, to achieve proper initial vapor distribution
- Possible restrictions to vapor flow, from the packing grid support, or holddown
- Problems in inspecting the final installation

Advantages of Packing vs. Trays

Often, we remove heat from a tower, at an intermediate point, by use of a *pumparound* or circulating reflux. Figure 7.11 is a sketch of such a pumparound. In many towers, the liquid flows in the pumparound section are greater than in the other sections, which are used for fractionation. That is why we are often short of capacity and initiate flooding in the pumparound or heat-removal section of a column.

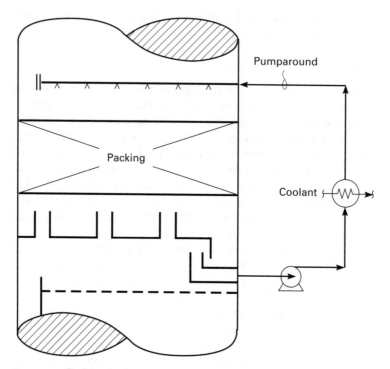

Figure 7.11 Packing is often used in pumparound sections of trayed towers.

Since we do not rely on pumparounds to fractionate—but just to remove heat—good vapor-liquid distribution is not critical. A bed of 4 or 5 ft of structured packing is often, then, an excellent selection for the pumparound section of a tower. The capacity of such a bed potentially has a 30 to 40 percent advantage over trays.

Once a tower's diameter is 3 ft or less, trays become difficult to install, due to the confined working area. Also, the tray support ring appreciably cuts down on the tower's cross-sectional area. Dumped-type packings (rings, saddles, broken beer bottles) do not present such problems. Also, for high liquid flows in a smaller-diameter tower (less than 5 ft), a well-designed packed tower will have a very substantial capacity advantage over a trayed tower. In general, the lower the pressure of a tower, the greater the advantage of using packing over trays.

Finally, the pressure drop of the vapor flowing through a packed tower will be an order of magnitude less than through a trayed tower. For vacuum distillation service, this is often of critical importance.

8

Steam and Condensate Systems

Water Hammer and Condensate Backup Steam-Side Reboiler Control

In this chapter we discuss process equipment such as

- The steam side of reboilers
- Steam turbine surface condensers
- Condensate recovery systems
- Deaerators
- The steam side of shell-and-tube steam preheaters

A few of the nasty features of this sort of process equipment we discuss are

- Condensate backup
- Accumulation of carbon dioxide
- Steam hammer
- Blown condensate seals

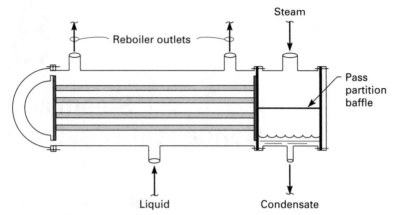

Figure 8.1   A shell and tube steam reboiler.

Steam Reboilers

When considering the steam side of steam heated reboilers, it is best to think about the reboiler as a steam condenser. The steam, at least for a conventional horizontal reboiler, is usually on the tube side of the exchanger, as shown in Fig. 8.1. The steam is on the tube side, because the shell side was selected for the process fluid. If the reboiler is a thermosyphon, or natural-circulation reboiler, then low-process-side pressure drop is important. For a horizontal reboiler, it is easiest to obtain a low pressure drop for the fluid being vaporized by placing it on the shell side.

The steam enters through the top of the channel head of the reboiler. Any superheat in the steam is quickly lost to the tubes. Superheated steam does very little in increasing heat-transfer rates in a reboiler. Actually, when considering the temperature difference between the steam and the process fluid, it is best to use the *saturated steam* temperature, as the real temperature at which all the heat in the steam, is available. For example, assume the following steam flow to a reboiler:

- 1000 lb/h of 100-psig steam
- 400°F steam inlet temperature
- 300°F condensate outlet temperature

The 100-psig steam condenses at approximately 320°F. The heat available from the steam is

- Desuperheating = (400°F − 320°F) × [0.5 Btu/(lb)(°F)] × 1000 lb/h = 40,000 Btu/h
- Condensing = 900 Btu/lb × 1000 lb/h = 900,000 Btu/h

- Subcooling = (320°F − 300°F) × [1.0 Btu/(lb)(°F)] × 1000 lb/h = 20,000 Btu/h
- Total reboiler duty = 960,000 Btu/h

This calculation is typical, in that 94% of the heat is liberated at the 320°F condensing temperature of the saturated steam. Another way of stating the same idea is that a steam reboiler depends on *latent-heat transfer,* and not on *sensible-heat transfer.*

Condensing Heat-Transfer Rates

When a vapor condenses to a liquid, we say that the latent heat of condensation of the vapor is liberated. In a steam reboiler, this liberated heat is used to reboil the distillation tower. When a vapor, or more commonly a liquid, cools, we say that its sensible heat is reduced. For a small or slight temperature change, the change in latent heat might be large, while the change in sensible heat will be very small.

Heat exchange provided by sensible-heat transfer is improved when velocities are higher. Especially when the heating fluid is on the tube side of an exchanger, sensible-heat-transfer rates are always increased by high velocity.

Heat exchange provided by latent-heat transfer is improved when velocities are lower. It is my experience that this loss of heat transfer, at high velocity, is quite large when steam is flowing through the tube side of an exchanger. Theoretically, this happens because the condensing film of steam is blown off the tube surface by the high vapor velocity.

This improved heat-transfer rate, promoted by low velocity, applies not only for condensing steam but also for condensing other pure-component vapors. And since condensation rates are favored by low velocity, this permits the engineer to design the steam side of reboilers and condensers in general, for low-pressure drops. For example, if we measured the pressure above the *channel head pass partition* baffle shown in Fig. 8.1, we would observe a pressure of 100 psig. The pressure below the channel head pass partition baffle would typically be 99 psig.

Blown condensate seal

Figure 8.2 shows a common type of reboiler failure. The steam trap on the condensate drain line has stuck open. A *steam trap* is a device intended to open when its float is lifted by water. The steam trap remains open until all the water drains out of the trap. Then, when there is no more water to keep the trap open, it shuts. But, if the float sticks open, steam can blow through the steam trap. This is called a *blown condensate seal.* The average vapor velocity through the tubes

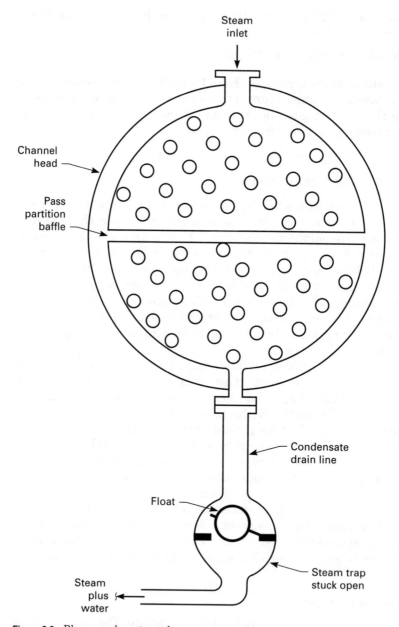

Figure 8.2 Blown condensate seal.

will then accelerate. If the steam trap is passing a lot of steam, the velocity of steam in the tubes increases a bit. I have seen this often. Blowing the condensate seal, due to a faulty steam trap, causes a loss in reboiler duty of 50 percent or more. This is due to the increased steam tube-side velocity, reducing the rate of condensation of the steam, and hence reducing the rate of liberation of the latent heat of condensation of the steam.

Condensate backup

What would happen to a steam reboiler if the float in the steam trap became stuck in a partly closed position, or if the steam trap were too small? Water—that is, steam condensate—would start to back up into the channel head of the reboiler, as shown in Fig. 8.3. The bottom tubes of the reboiler bundle would become covered with water. The number of tubes exposed to the condensing steam would decrease. This would reduce the rate of steam condensation, and also the reboiler heat duty.

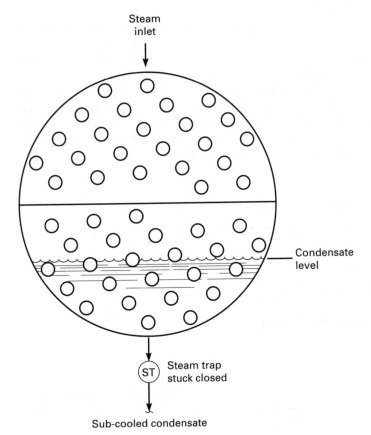

Figure 8.3   Effect of condensate backup.

Meanwhile, the tubes covered with stagnant water would begin to cool. The steam condensate itself around these tubes would cool. This cooled water would be colder than the saturation temperature of the condensing steam. The tubes would then be said to be submerged in *subcooled* water.

We can see, then, that either condensate backup, or blowing the condensate seal, will cause a steam reboiler to lose capacity. If you think either of these two problems could cause a loss in reboiler duty, try opening the bypass around the steam trap. If the reboiler duty goes up, the problem was condensate backup. If the reboiler duty goes down, then the problem might be a blown condensate seal. If it looks like a blown condensate seal problem, close the steam trap bypass. Then, partially close the valve downstream of the steam trap. If this increases the reboiler duty, a blown condensate seal failure is proved.

Maintaining System Efficiency

Steam flow control

The flow of steam to a reboiler can be controlled by using a control valve on either (1) the *steam inlet line* or (2) the *condensate outlet line*.

Figure 8.4 shows a control valve on the steam inlet line. The rate of steam flow to the reboiler is not really controlled directly, however, by this control valve. The actual rate of steam flow to the reboiler is controlled by the rate of condensation of the steam inside the tubes. The faster the steam condenses, the faster it flows into the channel head. The function of the control valve is to reduce the steam pressure in the channel head of the reboiler. For example, in case 1:

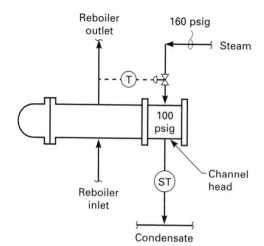

Figure 8.4 Varying channel head pressure controls heat input.

- 160 psig steam supply header pressure.
- ΔP across control valve = 60 psi and the control valve is 40 percent open.
- 100 psig steam pressure, condensing at 320°F, in the channel head.
- Shell-side reboiler temperature of 240°F.
- Steam flow = 10,000 lb/h
- Temperature difference between the condensing steam and the boiling process liquid is then (320°F − 240°F) = 80°F. This is called the *temperature-difference* driving force, or ΔT.

Now the rate or heat duty of steam condensation is termed Q:

$$Q = A \times U(\Delta T)$$

where A = surface area of the tubes, which are exposed to the condensing steam, ft^2

U = heat-transfer coefficient—a constant, describing the rate of condensation of steam, on those tubes exposed to the condensing steam, (Btu/[(h)/(ft^2)(°F)]

ΔT = temperature difference between the shell and tube sides, °F

Q = heat-exchanger duty, Btu/h (for the preceding equation in case 1, 10,000,000 Btu/h).

In case 2, the shell-side reboiler temperature rises from 240 to 280°F (one reason for such a rise in temperature could be an increase in tower pressure). Now ΔT = (320°F − 280°F) = 40°F. Looking at the equation above, it looks as if Q will drop in half to 5000 lb/h (which is about the same as 5,000,000 Btu/h). Thus, the flow of steam to the reboiler has been cut in half, even though the control valve position has not moved.

You might have noticed how we have used 1,000,000 Btu/h of heat, interchangeably with 1000 lb/h of steam. This is approximately correct, for low-pressure steam.

To consider a third case, we wish to maintain the original 240°F shell-side temperature, but to increase the steam flow from 10,000 to 15,000 lb/h. This will force the steam inlet control valve to open. As the control valve opens, the pressure in the channel head rises from 100 psig to the full steam header pressure of 160 psig. At this pressure, steam condenses at 360°F. The new ΔT is then (360°F − 240°F) = 120°F. This new temperature driving force is 50 percent greater than the case one driving force of 80 percent. Hence the rate of steam condensation also increases by 50 percent, from 10,000 to 15,000 lb/h.

In case 1, the steam inlet control valve was 40 percent open. In case 3, let's assume that the inlet control valve is 70 percent open. If we open the control valve to 100 percent, the steam flow will not increase at all. Why? Because, once the steam pressure in the channel head rises to the steam header pressure, no further increase in steam flow is possible, regardless of the position of the inlet control valve.

Condensate control

Once the steam pressure in the channel head of Fig. 8.4 falls to the pressure in the condensate collection header, the steam trap can no longer pass condensate. Water will back up in the channel head, and water-log the lower tubes in the tube bundle. This will lead to unstable steam flow control. This is especially true if the steam supply pressure is less than 20 psig higher than the maximum condensate collection header pressure.

It is better not to use a steam inlet control valve when using low-pressure steam. The channel head pressure will then always equal the steam header supply pressure. The flow of steam to the reboiler can then be controlled only by raising or lowering the water level in the channel head, as shown in Fig. 8.5. This sort of control scheme will work perfectly well until the water level drops to the bottom of the channel head. If the condensate drain control valve then opens further, in an attempt to increase steam flow into the reboiler, the condensate seal is blown, and the reboiler heat duty drops.

A better design is shown in Fig. 8.6. In this scheme, a condensate drum is used to monitor the level in the channel head. As the drum level is drawn down, the number of tubes in the reboiler exposed to the condensing steam is increased. However, when the water level drops to

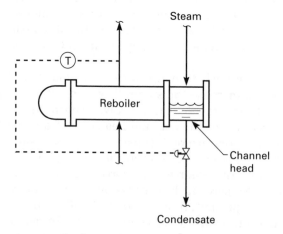

Figure 8.5 Condensate backup controls steam flow.

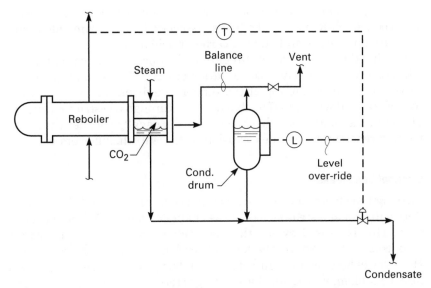

Figure 8.6 Venting below the pass partition baffle stops carbonic acid corrosion.

the bottom of the channel head, the level sensor in the condensate drum assumes control (i.e., overrides) over the condensate flow controller, and prevents loss of the condensate seal.

One important feature of Fig. 8.6 is the *condensate drum balance line*. Note, that this line is connected below the channel head pass partition baffle. This ensures that the pressure in the channel head, below the pass partition baffle, and the pressure in the condensate drum, are the same. If these two pressures are not identical, then the level in the condensate drum cannot represent the level in the channel head. For this reason, never connect the condensate drum vapor space to either the steam supply line or the top vent of the reboiler's channel head.

Carbonic Acid Corrosion

Steam produced from demineralized water is free of carbonates. Steam produced from lime-softened water will be contaminated with carbonates that decompose in the boiler to carbon dioxide. As the steam condenses in a reboiler, the CO_2 accumulates as a noncondensable gas. This gas will be trapped mainly below the channel head pass partition baffle shown in Fig. 8.6. As the concentration of CO_2 increases, the CO_2 will be forced to dissolve in the water:

$$H_2O + CO_2 = H_2CO_3 = (H^+)(HCO_3^-)$$

That is, carbonic acid, will be formed. Carbonic acid is quite corrosive to carbon steel. Reboiler tube leaks, associated with steam-side corrosion, are almost certainly due to carbonic acid attack.

Venting the channel head through the balance line shown in Fig. 8.6 will prevent an excessive accumulation of CO_2. This is done by continuous venting from the top of the condensate drum. For every 10,000 lb/h of steam flow, vent off 50 lb/h of vapor through a restriction orifice, placed in the condensate drum vent. This is usually cheaper than controlling reboiler steam-side corrosion, with neutralizing chemicals.

Channel head leaks

Varying the steam-to-condensate interface level to control the reboiler duty will promote steam leaks in the channel head-to-shell flanged closure. This is caused by the thermal cycling and stresses that result from constantly varying the level of condensate in the channel head. However, when low-pressure steam (<60 psig) is used, this becomes a minor problem, which may be safely ignored.

When high-pressure steam (>100 psig) is used, rather significant leaks of hot condensate and steam can be caused by a variable condensate level in the channel head. For such higher-pressure steam sources, control of steam flow with condensate backup, as shown in Figs. 8.5 or 8.6, is best avoided.

Condensate Collection Systems

How much steam condensate is your plant recovering? Seventy percent is considered pretty good, and 30 percent is, by any standard, pretty awful. As condensate collection flows are rarely metered, here is a really good way to make such an overall measurement (ST = steam; TR = treated water):

- Determine ST—the lb/h of steam raised in the whole plant
- Determine TR—the lb/h of softened or demineralized treated water flowing to the deaerators
- The percent condensate recovery is then

$$100\% - 100\% \,(TR \div ST)$$

Loss of steam condensate to the plant's sewer is environmentally wrong, and wastes money for water-treating chemicals and energy. The principal reasons why steam condensate is lost to the sewer are

- It creates steam or water hammer in the condensate collection system.
- Backpressure from the condensate collection lines creates control difficulties in steam reboilers or heaters.
- The condensate is contaminated with traces of dissolved hydrocarbons, phenols, NH_3, H_2S, etc.

Water hammer

Steam, or water hammer (more properly called *hydraulic hammer*), is one process plant phenomenon familiar to the general public. I well remember trying to warm up the steam system of a large amine plant in Texas City in 1980, and feeling more than hearing, the crescendo of crashes, due to steam hammer. The cause of steam hammer is illustrated in Fig. 8.7.

In general, *hydraulic hammer* is caused by the sudden conversion of the velocity of a liquid into pressure, causing a surge of pressure inside a piping system. *Steam hammer* is caused by the creation of localized cool areas in piping systems containing saturated steam. For example, when first introduced into idled pipes, the steam condenses rapidly on encountering a length of cold piping. This creates slugs of water. The continuing rapid-but-localized-condensation of steam further downstream creates areas of low pressure, or even a partial vacuum. The slugs of water rush to these areas of low pressure. Or, more precisely, the pressure differences created by the localized condensation of the

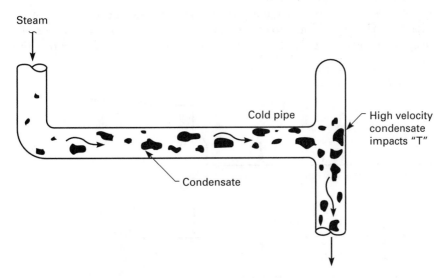

Figure 8.7 Water hammer.

steam provide a source of energy to accelerate water in the steam system. When these rapidly moving slugs of water impact an elbow or T-junction, the vibrations and noise of steam hammer result.

Introducing condensate from high-pressure steam traps into the low-pressure condensate collection system will generate steam. For example, passing 1000 lb/h of condensate from a 160-psig reboiler into a 20-psig collection system will generate about 100 lb/h of steam. More importantly, the volume of flow will increase from 20 ft³/h to over 1000 ft³/h, as a result of flash vaporization. If water at, say, 200°F from a low-pressure trap enters the piping, then any flashed steam will rapidly condense, and create an area of low pressure. Slugs of water will rush to this low-pressure area, crashing into elbows and piping fittings; thus, the origin of steam hammer.

Operators stop water hammer by dumping either the hot condensate (from a high-pressure steam trap) or the colder condensate (from a low-pressure steam trap) to the sewer. Either way, the condensate is no longer recycled to boiler feedwater. The sort of design required to collect condensate without steam hammer is illustrated in Fig. 8.8. Basically, hot, high-pressure condensate is collected in a dedicated steam. The flashed steam from this system is recovered as low-pressure steam. The resulting water is then passed into the low-pressure condensate recovery piping system, along with water flowing from the low-pressure steam traps.

Condensate backup in reboilers

Operators who have problems with loss of reboiler capacity often attribute these problems to condensate backup. This is usually true. To drop the level of water out of channel head, either the steam trap or the

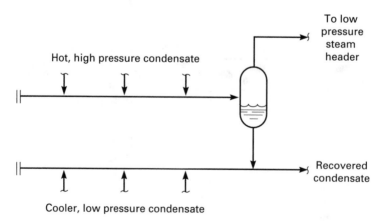

Figure 8.8 Avoiding water hammer in a condensate collection system.

condensate drum is bypassed, by putting the condensate to the sewer. Sometimes the float of the trap is sticking, but mostly, the difficulty is an erratically high pressure in the condensate collection piping. Our guess is that the engineers who design these collection systems do not anticipate the large volume of steam generated from flashing, high-pressure condensate.

Condensate pumps are sometimes used to overcome such backpressure problems. However, these pumps are often not kept in good repair, and condensate is still lost to the sewer. Eliminating the steam inlet control valve of the type shown in Fig. 8.4 has helped recover condensate from many reboilers, supplied with low-pressure steam.

Contaminated condensate reuse

Much of the steam consumed in process units comes into direct contact with the process streams. A few examples are

- Steam vacuum jets
- Catalyst lift steam
- Stripping steam
- Vessel purge steam

For corrosion and safety reasons, the condensate recovered from these sources is best not returned to the deaerator for use as boiler feedwater. However, depending on the contaminant, the condensate may be reused for a number of services. Our favorite reuse of such contaminated condensate is as a replacement for *velocity steam* in the heater-tube passes of a fired furnace.

There is little danger, in injecting a controlled amount of water into a furnace inlet, when using a properly designed *metering pump*. Such pumps typically have a capacity of 1 to 10 GPM and provide a set flow, regardless of the discharge pressure. The injected water flashes immediately to steam inside the furnace tubes. We have retrofitted several vacuum and delayed coker heaters with condensate injection systems, with no adverse downstream effects. Water from the hot well of a vacuum ejector system is our normal source of condensate for this environmentally friendly modification.

Deaerators

The humble deaerator, operated by the Utility Department, is an interesting and important component of any process facility. Oxygen is a highly corrosive element, and if left in the boiler feedwater, would rapidly oxidize the boiler's tubes.

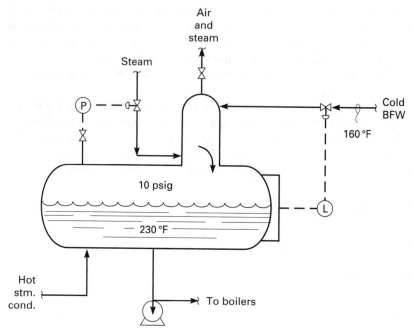

Figure 8.9 A deaerator.

The dissolved air left in boiler feedwater (BFW) is stripped out with steam, in the deaerator shown in Fig. 8.9. The cold BFW has been taken first from the Mississippi River and then filtered to remove sand and sediment. Removal of the bulk of the calcium salts that would cause hardness deposits in the boilers is often accomplished by hot-lime softening. If excess CO_2 gas appears in downstream units consuming the steam, it is the fault of the lime softening, not the deaerator.

Figure 8.9 shows that the cold BFW has been heated to 160°F, from the 90°F lime-softener effluent. This is an excellent way to save steam to the deaerator. Usually, heating river or well water above 130°F would cause the laydown of hardness deposits inside the cooling-water tubes (assuming the cooling water to be on the tube side). But using softened water as a cooling-water supply permits that water to be easily heated to 160°F, without fear of precipitating hardness deposits. Thus softened cooling water can also be used as boiler feedwater, and about 40 percent of the deaerator steam requirements are then saved.

Most of the steam supply to a deaerator is used to heat the 160°F BFW to 230°F. This is the boiling point of water at 10 psig, which is the pressure in the deaerator. The water in the deaerator must always be at its boiling point; it is impossible to steam-strip air out of water, below its boiling-point temperature.

The 160°F BFW is efficiently mixed with the incoming steam, in what is effectively a small, vertical stripping tower, mounted above the large deaerator drum. The majority of the steam condenses by direct contact with the 160°F BFW and in so doing, the latent heat of condensation of the steam is used to increase the sensible-heat content of the 160°F BFW to 230°F.

A small amount of steam is vented from the top of the stripping tower to the atmosphere. Using a *gate valve,* with a hole drilled through the gate, is a simple way to control the venting rate. The dissolved air in the cold BFW is vented with this steam.

The stripped 230°F BFW drains into the large deaerator drum. This drum simply provides residence time for the high-pressure BFW pumps, which supply water directly to the boilers. Recovered steam condensate, which should be air-free, is fed to this drum through a separate nozzle.

Deaerator flooding

One of the most interesting troubleshooting assignments of my career involved a deaerator in New Orleans. The operator reported loss of the water level in the deaerator drum. The level-control valve shown in Fig. 8.9 would open 100 percent, but this drove the level down even faster. The operator reported that the only way he could restore the water level would be to mostly close the water makeup level-control valve.

On the surface, this story sounds crazy. But, let's see what happened. This deaerator had been designed for a much smaller flow of 160°F BFW, and a much larger flow of hot-steam condensate, than are current operations. The cold BFW feed line had been oversized, but the steam line was of marginal size. As the demand for hot BFW increased, the cold-BFW level-control valve opened. This reduced the temperature and pressure in the deaerator drum. In response, the steam pressure-control valve also opened. But when the cold-BFW level-control valve was 40 percent open, the steam pressure-control valve was 100 percent. Steam flow was now maxed out.

The pressure inside the deaerator started to drop, as there was not enough steam flow to keep the water in the drum at its boiling point. The reduction in the deaerator pressure increased the volume of steam flow through the bottom tray of the stripping tower. Why? According to the *natural-gas law:*

$$\text{Volume} \sim \frac{1}{\text{pressure}}$$

In words, this means that volume is inversely proportional to pressure. As the pressure of the flowing steam declined, the steam's volume increased. The larger volume of steam flow resulted in a higher vapor velocity to the stripper trays. This caused the bottom stripper tray to flood. After a few minutes, the flooding forced cold BFW out of the atmospheric vent. The level in the deaerator drum then fell. The cold BFW level-control valve opened further, driving down both the temperature and the pressure in the deaerator drum. The volume and the velocity of the steam flow to the stripper also increased, and the flooding became progressively more severe.

The only way the operator could restore the water level was to reverse this process. He manually restricted the flow of cold BFW through the level-control valve. This raised the stripper pressure, and stopped the flooding.

The problem was corrected by increasing the temperature of the cold BFW. Note that the key to solving this problem was observing the loss of water through the atmospheric vent. Sometimes, as far-fetched as an operator's description of a problem seems, it is still correct.

Surface Condensers

Steam turbine surface condensers

Steam used to drive a turbine can be extracted at an intermediate pressure, for further use of the low-pressure steam. Rarely is the steam vented to the atmosphere, as this wastes steam, and the condensate is also lost. Many turbines exhaust steam, under vacuum, to a surface condenser. The lower the pressure in the surface condenser, the greater the amount of work that can be extracted from each pound of steam (see Chap. 17).

Discounting the presence of air leaks, the temperature inside the surface condenser determines the pressure of the steam exhausting from the turbine. This pressure is the vapor pressure of water at the surface condenser outlet temperature.

The original steam condensers were *barometric condensers,* which were used to increase the efficiency of the steam-driven reciprocating beam engines by a factor of 10. The barometric condenser was invented by James Watt (the steam engine was invented by Thomas Newcomen). Exhaust steam is mixed directly with cold water. As this creates a vacuum, the barometric condenser must be elevated about 30 ft above grade. The mixed condensate and cooling water drains through a pipe called a *barometric leg*—hence the name *barometric condenser.*

The *surface condenser* is an improvement on the barometric condenser, because it permits recovery of clean steam condensate. Other than this factor, the old-fashioned barometric condenser is more efficient than the more modern surface condenser.

Figure 8.10 shows the type of surface condenser widely used on older steam turbines. Note that it has both a vapor and a liquid outlet. The

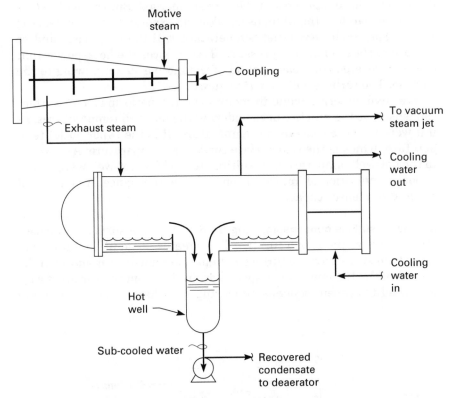

Figure 8.10 A surface condenser serving a steam turbine.

turbine is located above the surface condenser. The wet exhaust steam flows down into the top of the condenser shell. Note that the exhaust steam from an efficient turbine will contain several percent of water.

The shell-side pressure drop of the surface condenser is quite low. The vapor outlet flow consists of air drawn into the system through leaks, CO_2, and a small amount of uncondensed steam. The weight flow of vapor from the top of the condenser is 1 percent or less than the flow of condensate from the bottom of the condenser.

The vapor is drawn into a steam jet (discussed in Chap. 16). The steam condensate flows into the boot or *hot well*. The water in the boot is slightly subcooled. This is accomplished by a pair of baffles that create a small zone of condensate backup. The subcooled condensate, cooled to perhaps 10°F below its boiling or bubbling point, is easier to pump. As the pressure in the hot well is subatmospheric, the hot-well pump typically develops a ΔP of at least 30 to 50 psi.

If the hot-well pump cannot handle the required condensate flow, the water level in the well will back up. The temperature in the well will go down as the lower condenser tubes are submerged. But this will reduce the surface area of the condenser, which is exposed to the condensing steam. The condenser outlet temperature will therefore rise, and so will the surface condenser pressure. This reduces the horsepower that can be recovered from each pound of the motive steam flowing to the turbine. The turbine will then slow down.

One common error made in monitoring the performance of surface condensers is the practice of considering the hot-well temperatures, as if it were the true condensing temperature. It is the vapor outlet temperature, which is the real surface condenser temperature. A decrease in the hot-well temperature, resulting from a high hot-well water level, is not an indication of improved condenser performance. It is a sign of reduced condenser capacity.

Air-cooled surface condensers.   Figure 8.11 shows a surface condenser elevated above the steam turbine. This creates an additional problem, in that moisture from the turbine exhaust steam will accumulate in the bottom of the turbine case. A special drain line from the turbine's case is needed to prevent condensate backup from damaging the spinning wheels.

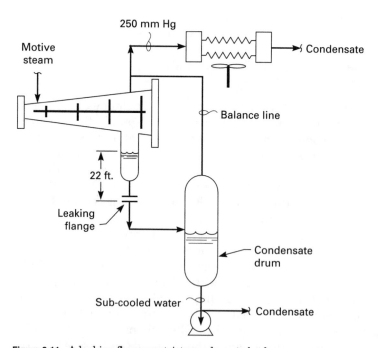

Figure 8.11   A leaking flange restricts condensate backup.

One such turbine, in a refinery near London, would not drain properly. In order to push the condensate out of the turbine case, the operators were forced to raise the surface condenser pressure from 100 to 250 mm Hg (i.e., 20 in of mercury vacuum, in the American system). Note that the balance line shown in Fig. 8.11 keeps the pressure in the turbine case and the condensate drum, into which the turbine case is draining, both equal at the same pressure.

The turbine case pressure was increased by raising the pressure in the air-cooled surface condenser. This was accomplished by shutting off several of the air fans, which, in turn, increased the condensing temperature of the exhaust steam. But why would raising the turbine case pressure drain the turbine, anyway? After all, increasing the surface condenser pressure also increased the pressure in the drum that the turbine case drained to.

The answer is revealed when the available data are converted to consistent units. The 250 mm Hg turbine case pressure is equal to a height of water of 11 ft. Atmospheric pressure in London was 14.5 psia that day, which is equal to 33 ft of water. The difference between atmospheric pressure and the pressure in the turbine case, expressed in feet of water, is then

$$(33 \text{ ft} - 11 \text{ ft}) = 22 \text{ ft}$$

This turned out to be the exact elevation difference between the water level in the turbine case, and the only flange on the drain line, from the turbine case to the drum below. This flange was found to be leaking. If the water head pressure at the flange became less than atmospheric pressure, then air was forced into the drain line. The bubbles of air expanded as they floated up the drain line into the turbine case. This prevented the water from draining freely through the drain line into the drum. Only by raising the turbine case pressure, to pressure-up the flange, to match atmospheric pressure could this air leak be stopped.

The flange leak was taped over, and the exhaust-steam pressure dropped back to 100 mm Hg. The steam required to drive the turbine fell by 18 percent. This incident is technically quite similar to losing the downcomer seal on a distillation tower tray. Again, it illustrates the sort of field observations one needs to combine with basic technical calculations. This is the optimum way to attack, and solve, process problems.

9

Bubble Point and Dew Point

Equilibrium Concepts in Vapor-Liquid Mixtures

Our work as process engineers and operators is based on three principles:

- *Hydraulics*—e.g., when the velocity of water in a pipe goes up, its pressure goes down.
- *Heat balance*—e.g., condensing steam gives up latent heat, or sensible heat, to increase the temperature in the feed preheater.
- *Vapor-liquid equilibrium*—e.g., a boiling liquid is at its bubble point, and a condensing vapor is at its dew point.

Bubble Point

The purpose of this chapter is to explain what is meant by the terms *bubble point* and *dew point,* and how we can use these ideas to improve the operation of the distillation tower. To begin, we will derive the *bubble-point equation,* from the basic statement of vapor-liquid equilibrium:

$$y_1 = K_1 \cdot x_1 \tag{9.1}$$

where y_1 = concentration of the first component in the vapor
   x_1 = concentration of the first component in the liquid
   K_1 = an equilibrium constant

We really should use *mole fraction,* and not concentration, in our description of y and x, but for our work, we will just say that the term *concentration* refers to the percent of a component that the operator would see in the gas-chromatographic (GC) results, as reported by the lab. The equilibrium constant, assuming the *ideal-gas law* applies, is defined as

$$K_1 = \frac{P_{V,1}}{P_T} \qquad (9.2)$$

where $P_{V,1}$ = vapor pressure of the first component, at the tempera-
      ture we are working at, in psia (see Fig. 9.1 for chart of
      vapor pressures used here)
   P_T = total pressure we are working at, in psia (psia = psig +
      14.7)

If you do not recall the meanings of *mole fraction* or the *ideal-gas law,* don't worry—it is not necessary to recall these in order to understand bubble points, or dew points. Substituting Eq. (9.2) in Eq. (9.1), we obtain

$$y_1 = \frac{P_{V,1} x_1}{P_T} \qquad (9.3)$$

Let's assume that we have three components in the vessel shown in Fig. 9.2. Then we could write

$$y_1 + y_2 + y_3 = \frac{P_{V,1} x_1}{P_T} + \frac{P_{V,2} x_2}{P_T} + \frac{P_{V,3} x_3}{P_T} \qquad (9.4)$$

But if we add up the concentration of the three components in the vapor phase on the left side of Eq. (9.4), we would get 100 percent. The fractions on the right side of Eq. (9.4) all have the same denominator (i.e., P_T), so they can be also added together:

$$100\% = \frac{P_{V,1} x_1 + P_{V,2} x_2 + P_{V,3} x_3}{P_T} \qquad (9.5)$$

Recalling that 100 percent of anything is the whole thing or, in other words, equals unity or one, if we cross-multiply both sides of this equation by P_T, we have

$$P_T = P_{V,1} x_1 + P_{V,2} x_2 + P_{V,3} x_3 \qquad (9.6)$$

This is the bubble-point equation for a three-component system.

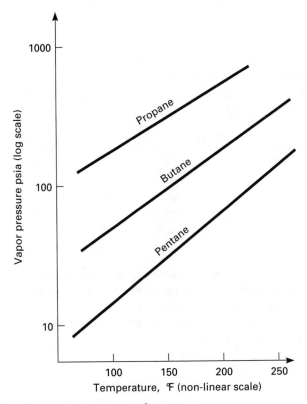

Figure 9.1 Vapor-pressure chart.

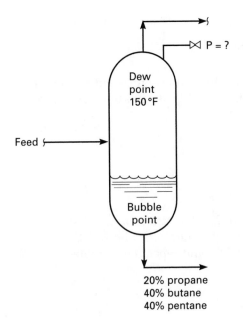

Figure 9.2 Calculating bubble-point pressure.

20% propane
40% butane
40% pentane

Using the bubble-point equation

Are we missing the pressure P_T in the flash drum shown in Fig. 9.2? Let's calculate this pressure, using the bubble-point equation and the vapor pressure chart shown in Fig. 9.1:

Component	Vapor pressure at 150°F, psia		Concentration in liquid, %		Partial pressure, psia
Propane	330	×	20	=	66
Butane	105	×	40	=	42
Pentane	35	×	40	=	14
Vessel pressure				=	122

The term *partial pressure,* meaning part of the total pressure created by each component, is important. The partial pressure of a component, divided by the total pressure, is the concentration of the component in the vapor phase. For example, the concentration of propane in the vapor leaving the drum shown in Fig. 9.2 is

$$\frac{66 \text{ psia}}{122 \text{ psia}} = 54\%$$

What is the concentration of pentane in the vapor? (Answer: $14 \div 122$ = 13 percent.)

Adjusting temperature to meet a product specification

A new set of product specifications has just been issued to your shift. The liquid from the flash drum shown in Fig. 9.3 has too much propane. The new liquid specification is

- Propane: 10 percent
- Butane: 40 percent
- Pentane: 50 percent

The pressure in the drum is still fixed at 122 psia. So, it seems as if we will have to run the drum hotter. But how *much* hotter? Suppose we raise the drum temperature to 160°F, and repeat our bubble-point calculation:

Component	Vapor pressure at 160°F, psia		Concentration in liquid, %		Partial pressure, psia
Propane	380	×	10	=	38
Butane	130	×	40	=	52
Pentane	40	×	50	=	20
Calculated vessel pressure				=	110

Apparently, our guess of 160°F was wrong. If we had guessed the correct temperature, the calculated vessel pressure would have been 122 psia, not the 110 psia. Try to work this problem yourself[1] by guessing a new flash drum temperature (answer: 168°F).

This seems to be a potentially good application for computer technology. For example, an operator is running a debutanizer, and finds that she has too much isobutane in her isopentane bottoms product. She enters the most recent gas chrome result from the lab in the computer, with the corresponding tower pressure, and reboiler outlet temperature. Next, she enters the isobutane specification she would like to achieve in the tower's bottoms product. The computer then tells her to raise the reboiler outlet temperature by 17°F, to get back on spec. (specification) quickly. This is a lot better than guessing at the reboiler temperature, lining out the tower, and waiting half the shift for the lab GC result, before making your next move.

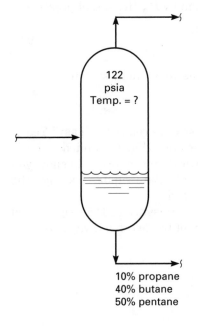

122 psia
Temp. = ?

10% propane
40% butane
50% pentane

Figure 9.3 Calculating a temperature to meet a new specification.

Dew Point

Dew-point calculations

We now derive the *dew-point equation* from the same basic statement of vapor-liquid equilibrium, starting with equation (9.3), in the previous section:

$$y_1 = \frac{P_{V,1}x_1}{P_T} \tag{9.3}$$

Now let's multiply both sides of this equation by $P_T/P_{V,1}$

$$\frac{y_1 P_T}{P_{V,1}} = x_1$$

Again, let's assume we have three components:

$$\frac{y_1 P_T}{P_{V,1}} + \frac{y_2 P_T}{P_{V,2}} + \frac{y_3 P_T}{P_{V,3}} = x_1 + x_2 + x_3 \tag{9.7}$$

However, if we add up the concentration of the three components in the liquid phase on the right-hand side of Equation (9.7), we would get 100 percent, which is unity or equal to one as before:

$$\frac{y_1 P_T}{P_{V,1}} + \frac{y_2 P_T}{P_{V,2}} + \frac{y_3 P_T}{P_{V,3}} = 1.0 \tag{9.8}$$

Next, we divide both sides of the equation by P_T, the vessel pressure:

$$\frac{y_1}{P_{V,1}} + \frac{y_2}{P_{V,2}} + \frac{y_3}{P_{V,3}} = \frac{1}{P_T} \tag{9.9}$$

This is the dew-point equation for a three-component system.

Using the dew-point equation

"Such simple maths," I hear you say. Is science really this easy? Yes, it certainly can be. In fact, much of the science that is applied to basic process engineering is very straightforward, and in order to show you how to calculate the temperature of the vapor leaving the depropanizer, in Fig. 9.4, that is all we need.

This time, we know that the tower-top pressure is 175 psig, or 190 psia. We also know that the composition of the overhead vapor is

- Propane: 80 percent
- Butane: 15 percent
- Pentane: 5 percent

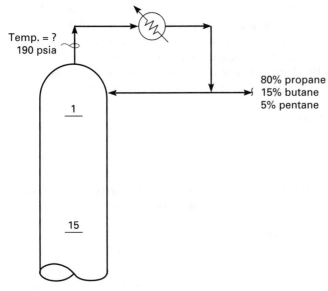

Figure 9.4 Calculating a dew-point temperature.

It is normal to assume that the vapor leaving the top of a tower is at its dew point. That is, it is at equilibrium with the liquid on the top tray of the tower. Unfortunately, this assumption falls apart if the tower is flooding and liquid is being entrained overhead from the column, with the vapor. However, assuming a normal, nonflooded condition, we will guess that the tower-top temperature is 140°F. Using the vapor-pressure curves provided in Fig. 9.1, we would calculate as follows:

Component	Vapor pressure at 140°F, psia	Concentration in vapor y	$y \div P_V$
Propane	300	0.80	0.00267
Butane	90	0.15	0.00166
Pentane	30	0.05	0.00167
Sum of quotients $y \div P_V$			0.00600

According to Eq. (9.9), the sum of the quotient of $y \div P_v$ ought to equal $1/P_T$:

$$0.00600 = \frac{1}{P_T}$$

Solving for P_T, we find the calculated tower-top pressure equals 167 psia. But the actual tower-top pressure is 190 psia. Evidently, we guessed too low a temperature. Try the calculation again yourself with a better guess for the tower-top temperature (answer: 146°F).

For this tower, the composition of the overhead vapor is the same as the overhead liquid product, made from the reflux accumulator.

It seems that the depropanizer overhead composition specification has been changed. Our new operating orders are to produce

- Propane: 90 percent

- Butane: 8 percent

- Pentane: 2 percent

The tower-top pressure is still 190 psia. The tower-top temperature will have to be reduced. Let's guess that it will be reduced to 130°F:

Component	Vapor pressure at 130°F, psia	Concentration in vapor y	$y \div P_V$
Propane	270	0.90	0.00333
Butane	80	0.08	0.00100
Pentane	26	0.02	0.00077
Sum of quotients $y \div P_V$			0.00510

Referring again to Eq. (9.9), we can solve for P_T, the tower pressure:

$$0.00510 = \frac{1}{P_T}$$

Or $P_T = 196$ psia. But this is the calculated tower pressure. The actual tower pressure is only 190 psia. Try to repeat this calculation to get the correct tower-top temperature (answer: 128°F).

Again, this seems to be a rather nice application for computer technology. Even a good-quality programmable calculator can store a number of vapor-pressure curves. At least for hydrocarbons, equations for these curves can be extracted from the API (American Petroleum Institute) data book. Also, a programmable calculator can perform bubble-point and dew-point calculations, with over 10 components, without difficulty.

Reference

1. American Petroleum Institute, *API Technical Data Book,* vol. I, sec. 5, "Vapor Pressures," Aug. 1964, fig. 5A 1.1, p. 5-3.

10

Steam Strippers

Source of Latent Heat of Vaporization

The use of steam to remove lower-boiling or lighter components from a liquid is one of the oldest methods of distillation. Sometimes called *steam distillation,* this technique relies on a combination of two simple effects.

Heat of Evaporation

The first effect is illustrated when we blow across a bowl of hot soup, to cool the soup. Our breath displaces the steam vapors that are on top of the soup. This encourages more molecules of steam vapors to escape from the soup; that is, the vapor pressure of the steam above the liquid soup is diminished, because steam is pushed out of the soup bowl with air. The correct technical way to express this idea is to say, "The partial pressure of the steam, in equilibrium with the soup, is diminished."

But our breath itself does not remove heat from the soup. The evaporation of steam from the soup, promoted by our breath, takes heat. Converting one pound of soup to one pound of steam requires 1000 Btu. This heat of evaporation comes not from our breath, but from the soup itself. The correct technical way to express this second effect is, "The sensible-heat content of the soup is converted to latent heat of evaporation."

Example calculations

For example, if we have 101 lb of soup in a rather large bowl, and cause one pound to evaporate by blowing across the bowl, the soup will lose 1000 Btu. This heat of evaporation will come at the expense of the temperature of the remaining soup in the bowl; that is, each pound of soup will lose 10 Btu. If the specific heat of our soup is 1.0 But/[(lb)(°F)], the soup will cool off by 10°F.

A steam stripper, as shown in Fig. 10.1, works in the same way. The diesel-oil product drawn from the fractionator column is contaminated with gasoline. The stripping steam mixes with the diesel-oil product on the trays inside the stripper tower. The steam reduces the hydrocarbon partial pressure and thus allows more gasoline to vaporize and to escape from the liquid phase into the vapor phase. The heat of vaporization of the gasoline cannot come from the steam, because the steam (at 300°F) is colder than the diesel oil (at 500°F). The heat of vaporization must come from the diesel-oil product itself.

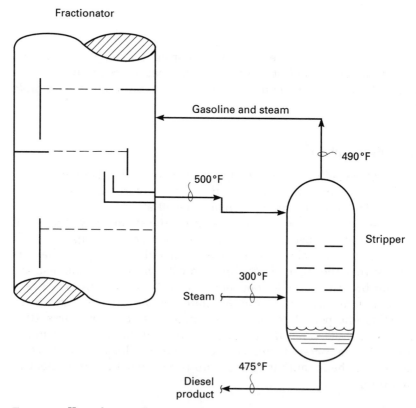

Figure 10.1 Heat of evaporation comes from product, not from steam.

We can use this idea to calculate the percent of diesel oil that would actually vaporize across the stripping trays in the stripping tower. Let's assume the following thermal properties for a typical hydrocarbon mixture of diesel and gasoline:

- 0.60 Btu/[(lb)(°F)] is the specific heat.
- 100 Btu/lb is the latent heat.

Referring to Fig. 10.1, the reduction in sensible heat of the diesel product equals:

$$(500°F - 475°F) \times 0.60 = 15 \text{ Btu/lb}$$

The percent of the feed to the stripper that evaporates is then

$$(15 \text{ Btu/lb}) \div 100 \text{ Btu/lb} = 15\%$$

Measuring evaporation in the field

Note that we have neglected the heat picked up by the steam in the preceding calculation. Often, the steam flow is quite small, compared to the stripper feed, so usually this effect may be disregarded. Unfortunately, we seldom may neglect ambient-heat losses.

Certainly, if the stripper tower and associated piping are radiating heat from the product, this is not contributing to stripping. To determine the temperature drop due to ambient-heat losses, proceed as follows:

- Check the temperature of the stripper feed.
- Shut off the stripping steam and wait 20 min.
- Check the temperature of the stripper bottoms product.

The difference between the two temperatures represents the ambient-heat loss associated with the stripper. Of course, with no ambient-heat loss, this temperature difference would be zero.

Stripper Efficiency

Many side-stream steam strippers, of the type shown in Fig. 10.1, do not work very well. Operating personnel report that the stripping steam is not effective in removing undesirable lighter components from the stripper feed. Why could this be so?

One of the main reasons for this sort of poor stripping efficiency is subcooled liquid feed to the stripper. Liquid drawn from any tower or vessel is assumed to be in equilibrium with the vapor phase in the tower or vessel. We say that the liquid is at its bubble point, or boiling point. We say that the vapor in the vessel is at its dew point, or saturation temperature.

When steam is mixed with a liquid at its bubble point, the partial pressure of the vapor, in contact with the liquid, is reduced. The liquid then begins to boil. The lighter components of the liquid are turned into vapor, and are carried out of the stripper, with the steam.

If liquid drawn from a column cools below its bubble point, as a result of ambient-heat loss, we say it is *subcooled*. Mixing a small amount of steam with subcooled liquid, will reduce the partial pressure of any vapor in contact with the liquid, but not enough to promote boiling. Eventually, as more and more steam is mixed with a subcooled liquid, it will begin to boil. But, for a given amount of steam, the amount of vapor that can be boiled out of a liquid will always be less if the liquid is subcooled. In this way, ambient-heat loss reduces the stripping efficiency of steam.

Wet steam will also reduce stripping efficiency. The water in the steam will be turned into steam when it contacts the hot diesel oil in the stripper, shown in Fig. 10.1. The heat of vaporization for this water must come from the sensible heat of the diesel. This reduces the temperature of the diesel, which also reduces its vapor pressure, which then makes it more difficult to vaporize its lighter gasoline components.

I was once working in a refinery that could not meet the flash-point specification for its diesel product. *Flash point* is the temperature at which a hydrocarbon will ignite, when exposed to an open flame. To raise the flash point of diesel oil, it is steam-stripped, to remove the lighter, more combustible components. I noticed that I could drain water from the bottom of the steam supply line to the diesel-oil stripper. I then screwed a *steam trap,* on to the ¾-in drain valve, on the steam supply line. The stripper bottoms temperature increased by 35°F, and the flash temperature of the diesel product increased from 120 to 175°F.

High liquid levels in the bottom of the stripper will also reduce stripping efficiency. A liquid level above the steam inlet will cause the stripping trays to flood. Flooding vastly decreases tray efficiency, and hence stripping efficiency.

As discussed in Chap. 2, tray deck dumping also greatly reduces tray efficiency. Unfortunately, steam strippers can have widely varying vapor rates, between the top and bottom trays of a column.

Vapor distribution in steam strippers

Figure 10.2 shows the type of hydrocarbon stripper discussed above. Note that the vapor load to the bottom tray is only the 3000 lb/h of stripping steam. The vapor load from the top tray is 15,000 lb/h. In other words, the vapor leaving the top tray of the stripper consists of 3000 lb/h of stripping steam *plus* 12,000 lb/h of hydrocarbon vapor.

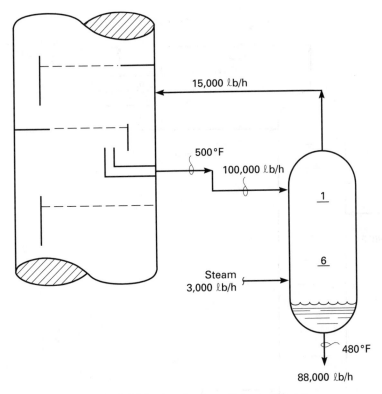

Figure 10.2 Vapor loads highest on top tray of steam stripper.

From the designer's point of view, the top tray of the stripper must have a several times greater number of sieve holes or valve caps on its tray deck than the bottom tray. If, however, all the trays in the stripper are identical, then either the bottom tray will leak (see Chap. 2), or the top tray will flood. Either way, stripping efficiency will suffer.

Steam stripping water

So far, we have been discussing the stripping of hydrocarbons. But of equal importance, is steam stripping of aqueous streams such as

- Removing benzene from seawater
- Removing ammonia from sour water
- Removing hydrogen sulfide from amine solutions
- Removing methanol from washwater
- Removing oxygen from boiler feedwater in deaerators

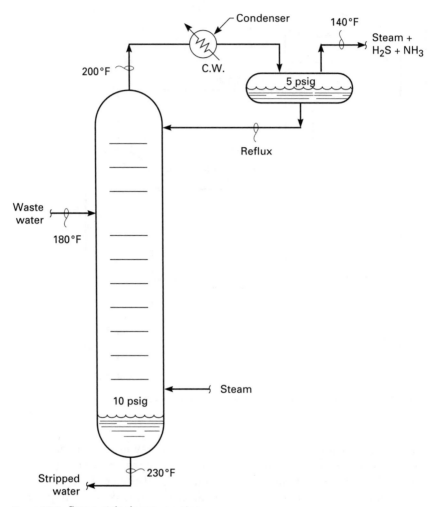

Figure 10.3 Steam stripping sour water.

Figure 10.3 shows a simple sour-water stripper. The steam is used to remove NH_3 and H_2S, dissolved in the waste, or sour water. In the diesel oil stripper discussed above, all the stripping steam went out the top of the stripper. But what happens to the stripping steam in a water stripper? It is used in four ways:

1. A large portion of the stripping steam is used to heat the waste-water feed, shown in Fig. 10.3, from 180 to 230°F. Almost all of this heat comes from the latent of condensation of the stripping steam, and very little of this heat comes from the temperature (or the sensible-heat

content) of the stripping steam. This means that the wastewater is heated by 50°F, by condensing steam inside the stripper.

2. A portion of the stripping steam is used to break the chemical bond between the water and the H_2S and NH_3, in the wastewater. When these gases first dissolved in the water, heat was evolved or released. This is called the *heat of solution*. When these same gases are driven out of the wastewater with the stripping steam, this same heat of solution has to be supplied. Again, this heat comes from condensing the stripping steam in the water flowing across the trays in the tower.

3. Some of the stripping steam condenses in the overhead condenser, shown in Fig. 10.3. The condensed steam, which accumulates in the reflux drum, is totally refluxed back to the top tray of the stripper tower.

4. A small amount of the stripping steam remains as steam, and leaves the reflux drum, with the H_2S and NH_3 vapor product.

From this distribution of the stripping steam, we can conclude that the pounds of vapor to the bottom tray is much larger than the pounds of vapor leaving the top tray. This is just the opposite of the diesel-oil stripper.

Temperature distribution in water strippers

The 230°F stripper bottom temperature, shown in Fig. 10.3, is simply the boiling point of water, at the 10-psig tower-bottom pressure. Small amounts of extraneous chemicals (phenol, alcohols, aromatics), dissolved in the stripped water, do not change the boiling-point temperature of this water.

The 200°F stripper tower-top temperature is the dew point of the vapors leaving the top tray. Most of these vapors are steam, and that is why the tower-top temperature is so high. The high steam content of the overhead vapors causes a water stripper to behave in a strange way! When the top reflux rate is increased, the tower-top temperature goes up, not down. This odd behavior is easily understood if we note that there is no liquid product made from the reflux drum. Therefore, the only way to increase the reflux rate, without losing the level in the reflux drum, is to increase the steam rate to the bottom of the stripper. The extra stripping steam drives up the tower-top temperature.

The 140°F reflux drum temperature, shown in Fig. 10.3, is the dew point of the vapors leaving the reflux drum. Almost always, we would like to minimize this particular temperature. The lower the reflux drum temperature, the smaller the amount of steam in the off-gas. If the off-gas is H_2S and NH_3, flowing to a sulfur recovery plant, the steam carried into the sulfur plant reduces the sulfur plant's capacity and efficiency.

On the other hand, a low reflux drum temperature increases the solubility of H_2S and NH_3 in the reflux water. As the concentration of H_2S and NH_3 in the reflux increases, the stripper has to work harder, to keep these components out of the stripped water.

Reboiled water strippers

Many water strippers are initially designed with steam reboilers, rather than with open stripping steam. The amount of steam required is the same in either case. The great advantage of the reboiled stripper is that the steam condensate is recovered, and recycled back to the boilers. When open stripping steam is used, the steam condensate is added to the stripped water, thus increasing the plant's water effluent. Hence, the use of open stripping steam is environmentally unfriendly.

Fouling is the main reason for abandoning many stripper reboilers and substituting open stripping steam. It is easy to pipe up steam to the stripper. It is difficult to determine and control the cause of the fouling in the reboiler. But not to find and control the cause of the fouling is sloppy engineering and poor operation.

Stripping aromatics from wastewater

Removing benzene and other aromatic compounds from a plant's effluent water is an increasingly common environmental requirement. This is typically achieved with a steam stripper. There is a rather neat trick, which can increase the stripper's efficiency: adding saltwater to the stripper feed. Aromatics, especially benzene, are far less soluble in brine than they are in freshwater. But, of course, the brine will be more corrosive than salt-free freshwater.

Side-stream stripper hydraulics

One of my first successful projects as a young engineer was to increase the quality of jet fuel from a side-stream stripper. Initially, I had thought the problem was due to poor stripping efficiency, but it was really a simple problem in *hydraulics:* the part of technology that deals with the pressure drops of gases and liquids flowing through pipes and other process equipment. Liquid heads, gravity, and friction are all subjects in hydraulics.

The particular problem I encountered is illustrated in Fig. 10.4. The jet fuel product was steam-stripped to remove a lighter naphtha contaminant. But much naphtha was left in the jet fuel. Apparently, the packing in the stripper tower was not working properly. However, a discussion with the unit operator indicated that they were using very little stripping steam. Introduction of a normal amount of steam resulted in a loss of liquid level in the bottom of the stripper.

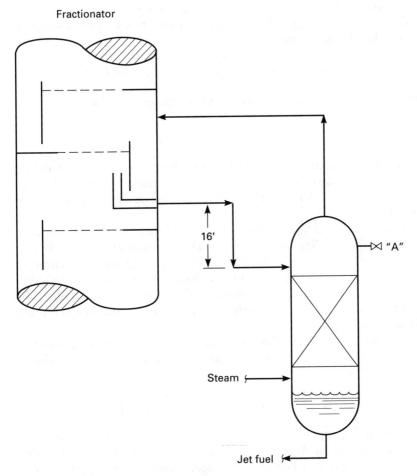

Fractionator

16'

⋈ "A"

Steam

Jet fuel

Figure 10.4 Hydraulics of a side-stream stripper.

The pressure at point "A" in Fig. 10.4 was 13 psig. This means that the pressure drop in the vapor line from the stripper, back to the fractionator, was 3 psig. In order for the unstripped jet fuel, to flow out of the lower-pressure fractionator, and into the higher-pressure stripper, it had to overcome this 3-psig pressure difference. The 16-ft elevation difference between the draw-off nozzle on the fractionator and the stripper inlet provided the necessary liquid head driving force.

Let us assume that the specific gravity of the unstripped jet fuel was 0.59. Also, note that for water, which has a specific gravity of 1.0:

$$1.0 \text{ psi of pressure} = 2.31 \text{ ft of water}$$

This means that the pressure head of a column of water 2.31 ft high, equals 1 psig. The height of a column of unstripped jet fuel, equal to a pressure head of one psig, is then

$$1.0 \text{ psi} = 2.31 \times \frac{1.0}{0.59} = 4.0 \text{ ft of jet fuel}$$

The pressure head of the column of unstripped jet fuel 16 ft high, shown in Fig. 10.4, is

$$16 \text{ ft} \div 4 \text{ ft/psi} = 4 \text{ psi}$$

This means that there was a 4-psi pressure head driving force available to overcome the 3-psi pressure drop of the stripper's overhead vapor line. This was sufficient for the jet fuel to flow out of the fractionator and into the stripper.

Raising the steam flow to the stripper increased the pressure drop in the overhead vapor line from 3 to 5 psi. The pressure at point "A" in Fig. 10.4 then increased from 13 to 15 psig. The 4-psi pressure head driving force was not sufficient to overcome the 5-psi pressure drop of the stripper's overhead vapor line. The unstripped jet fuel could no longer flow out of the fractionator and into the stripper, and the liquid level in the stripper was lost.

To correct this problem, I designed a larger-diameter vapor line connecting the stripper to the fractionator. Pressure drop through piping varies as follows:

$$\Delta P_{new} = \Delta P_{old} \left(\frac{\text{old pipe diameter}}{\text{new pipe diameter}} \right)^5$$

Changing the vapor line from a 3-in pipe to a 4-in pipe reduced the line's pressure drop from 3 to 0.7 psi. This permitted the stripping steam flow to be increased to the stripper, without impeding the jet fuel flow from the fractionator. The higher stripping steam flow efficiently removed the contaminant naphtha from the jet fuel product.

Liquid-line ΔP.

The liquid head driving force of 16 ft, or 4 psi, shown in Fig. 10.4, is actually not all available to overcome the higher stripper pressure. The frictional loss of the piping used to feed the stripper should be subtracted from the liquid head driving force. In the jet fuel example presented above, this frictional loss was neglected.

Sometimes the cheapest way to correct a hydraulic problem on a side-stream stripper is to pump the product into the stripper. I have modified several plants in this way, with excellent results.

11

Draw-off Nozzle Hydraulics

Nozzle Cavitation Due to Lack of Hydrostatic Head

I imagine that many readers might skip this chapter. After all, a nozzle is simply a hole in a vessel, flanged up to a pipe (see Fig. 11.1). Why a whole chapter? Well, it is not that simple.

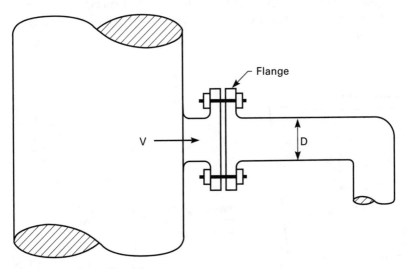

Figure 11.1 A draw-off nozzle.

Nozzle Exit Loss

The pressure drop of a fluid flowing through a nozzle is equal to

$$\Delta H = 0.34 \times V^2$$

where ΔH = pressure loss of the fluid as it flows through the nozzle, in inches of fluid
V = velocity of the fluid, as it flows through the nozzle, in feet per second

This equation assumes that before the fluid enters the nozzle, its velocity is small, compared to its velocity in the nozzle. The increase in the velocity, or the kinetic energy, of the fluid in the nozzle comes from the pressure of the fluid. This is Bernoulli's equation in action. The energy to accelerate the fluid in the draw-off nozzle comes from the potential energy of the fluid. This is Newton's second law of motion.

The coefficient used in the equation above (0.34) neglects friction and assumes the process fluid has a low viscosity. For most process nozzles, these are reasonable assumptions. Detailed information on draw-off nozzle coefficients has been published in Crane.[1]

Let's take a close look at the use of the term ΔH. Figure 11.2 shows a bucket with a hole punched in its side. The velocity of the water, as it escapes from the hole is V in the preceding equation. The height of water in the bucket, above the center of the hole, is ΔH. If we replaced the water in the bucket with gasoline, which has a lower density, the same hole velocity would occur with the same height of fluid.

The pressure drop of the fluid escaping from the bucket is called *nozzle exit loss*. Actually, nothing is lost; the potential energy, or pressure head of the water in the bucket, is just converted into velocity, or kinetic energy.

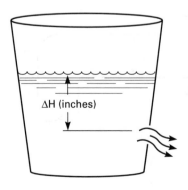

ΔH (inches)

Figure 11.2 "V" is the velocity of the water escaping from the hole.

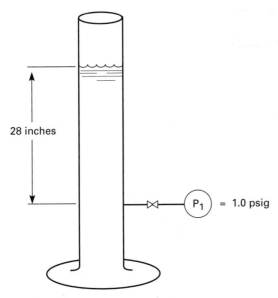

Figure 11.3 Water column exerting 1 psig of head pressure (1 psig = 28 in of water).

Converting ΔH to pressure drop

A column of water 28 inches high exerts a head pressure of 1 psi, as shown in Fig. 11.3. To determine the pressure drop of water flowing through a hole, in pounds per square inch, we would calculate

$$\Delta P = \frac{0.34}{28} \text{ in} \times V^2$$

where ΔP = pressure drop in psi.

But perhaps we have gasoline flowing through the hole. The density of water is 62 lb/ft^3; the density of gasoline is 40 lb/ft^3. Since the weight, or pressure head, of a column of fluid is proportional to its density, we calculate

$$\Delta P = \frac{0.34}{28 \text{ in}} \times \frac{40}{62} \times V^2$$

Again, ΔP is the pressure drop, in psi, of the gasoline escaping through a hole in the bucket. If we have a vapor flowing out of the overhead nozzle on a fractionator, as shown in Fig. 11.4, the pressure drop of the vapor escaping from the vessel in the overhead vapor line is

$$\Delta P = \frac{0.34}{28 \text{ in}} \times \frac{D_v}{62} \times V^2$$

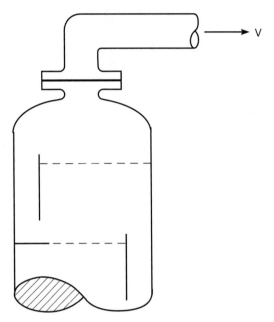

Figure 11.4   Nozzle exit loss for vapor flow.

where ΔP = pressure drop of a fluid, psi
$\quad D_v$ = density of a fluid such as a vapor, lb/ft³
$\quad V$ = velocity of fluid in nozzle, ft/s

This is an important equation to remember. It is the "pressure drop through a hole," equation. It works for water, steam, air, gasoline, alcohol, and all other fluids, unless viscosity is important. A viscosity of less than 20 or 30 centipoise (cp) (i.e., like hot maple syrup) means that viscosity is not important.

Critical Flow

Everything I have just said is wrong when the fluid we are working with undergoes *critical flow*. First, critical flow applies only to compressible fluids such as vapor, steam, air, or gases. Second, as long as the pressure drop that we calculate through the nozzle is less than 20 or 30 percent of the upstream pressure, we can ignore critical flow. Critical flow occurs when a compressible fluid velocity approaches the speed of sound: about 1000 ft/s. Process piping handling vapors is typically designed to work at a velocity of 100 ft/s.

Maintaining Nozzle Efficiency

Nozzle limitations

Figure 11.5 is an exact representation of my bathtub. When my tub has a water level of 28 in, the pressure right above the drain, at P_1, is 1 psi. Having bathed, I now pull the plug. I also start the water running into the tub, to keep the water level at 28 in.

The velocity of water flowing from the tub through the drain is 20 ft/s. The pressure drop, in psi, of the water as it escapes from the tub, is

$$\Delta P = \frac{0.34}{28 \text{ in}} \times (20)^2 = 5 \text{ psi}$$

The pressure at P_1 is now the 1-psi static head, minus the 5-psi nozzle exit loss, or negative 4 psig (or positive 10.7 psia). That is, the pressure at the drain is a substantial partial vacuum, or a *negative pressure,* meaning that it is below atmospheric pressure (atmospheric pressure at sea level is 14.7 psia).

This suggests that the pressure in a water drain can get so low, that air could be sucked out of the bathroom and down the drain. Of course, we all see this happen several times a day—typically when we flush a toilet. So much air is drawn into the water drainage piping, that we install vents on our roofs, to release this air. The only requirement, then, for vapors to be drawn into a flowing nozzle is for the nozzle exit loss to be larger than the static head of liquid above the nozzle.

Incidentally, if a bird builds its nest on top of one our roof toilet vents, we find the toilet will no longer flush properly. The experienced plumber states that the toilet won't flush because it is suffering from "vapor lock"; and this is true. A working knowledge of process equipment fundamentals often comes in quite handy around the home.

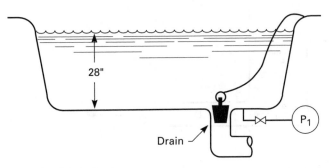

Figure 11.5 My bathtub.

Effect of bubble-point liquid

I have to confess to an odd habit. I sometimes bathe in boiling water. Draining a vessel full of liquid at its bubble, or boiling point prevents the formation of a negative pressure, or partial vacuum, in the drain. As soon as the vacuum begins to form, the liquid will begin to generate vapor. This vapor would tend to restrict the flow of liquid as it exits the drain, and thus reduce its flow rate. As the velocity of the liquid flowing into the drain decreases, the pressure of the liquid, at the inlet to the drain, will increase. This increase in pressure will be just sufficient to suppress vaporization of the liquid in the drain (i.e., the liquid velocity will be reduced just enough to cause a sufficient increase in pressure to prevent the liquid boiling).

Figure 11.6 illustrates the side draw-off of a fractionator. The liquid at point "A" is at its bubble or boiling point. We can be quite sure that this is true, because the liquid and vapor are in intimate contact at the vapor-liquid interface. We say that the vapor is at its dew point, and that the liquid is at its bubble point.

But how about the liquid at point "B"? Is this liquid also at its boiling or bubble point? It is the same liquid, having the same temperature and composition as the liquid at point "A." But the pressure at point "B" is slightly higher than the pressure at point "A."

The pressure at point "A" is the pressure in the vessel: 10 psig. The pressure at point "B" is the pressure at point "A" plus the static head of

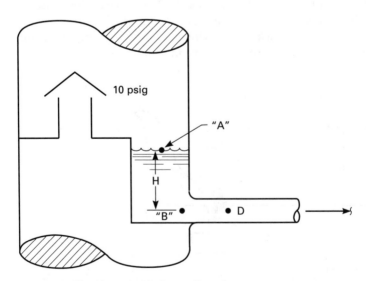

Figure 11.6 Fractionator side draw-off nozzle.

liquid (H), shown in Fig. 11.6. Let's assume that this liquid is water. Also, H equals 28 in. Then the pressure at point "B" is

$$10 \text{ psig} + \frac{28 \text{ in}}{28 \text{ in H}_2\text{O/psig}} = 11 \text{ psig}$$

The boiling point of water, at 10 psig, is about 240°F. This means that the temperature in the entire water draw-off sump is 240°F. But the pressure at point "B" is 1 psi above the water's bubble, or boiling point. We say, then, that the water at point "B" is subcooled by an equivalent of 28 in of liquid, or 1 psi.

But how about the pressure at point "D"? The liquid at both points "B" and "D," has the following in common:

- Same elevation
- Same fluid temperature
- Same fluid composition

If liquid is flowing through the nozzle, shown in Fig. 11.6, at, say, 9 ft/s, the pressure at point "D" will be lower than that at point "B." Assuming the velocity in the draw-off sump to be close to zero, we can calculate the pressure at point "D" as follows: head loss due to increased velocity = $0.34 \times 9^2 = 28$ in of water.

As 28 in of water equals 1 psi of head pressure, the head loss due to increased velocity at the nozzle exit equals 1 psi. Thus, the pressure at point "D" will then be 11 psig minus 1 psi, i.e., 10 psig.

Cavitation

For practice, calculate the pressure at point "D," assuming that the nozzle exit velocity is 10 ft/s. If you calculated 9.8 psig, you are likely wrong. You see, the water will start to flash to steam, if it falls below its boiling-point pressure. The boiling-point pressure of water at 240°F is 10 psig. As steam is evolved at a lower pressure, the large volume of vapor chokes off the flow in the draw-off nozzle. The flow of water slows down. As the velocity of the water decreases, its pressure increases. The pressure will increase to 10 psig—that is, the boiling-point pressure of water at 240°F. At this pressure, the vaporization of the liquid in the draw-off nozzle is zero.

In summary, the lowest pressure that can be reached at point "D" in Fig. 11.6 is the pressure at point "A." When these two pressures are equal, we say that the draw-off nozzle is limited by *cavitation*. If we were to lower the pressure downstream of point "D," say, by opening a control valve, the increase in flow would be zero.

External restrictions

Figure 11.7 shows two almost identical draw-off arrangements. The only difference is the elevation of the control valve in the draw-off line. Control valve "A" is at the same elevation as the draw-off nozzle. The pressure drop across the control valve is 2 psi, or 56 in of water. Let's assume that

■ Increase in velocity through the draw-off nozzle is small, and hence the nozzle exit loss is zero.

■ The frictional loss through the piping and nozzle is zero.

Even after we have made these two unlikely assumptions, the height of hot water in the draw-off sump must still be 56 in above the centerline of the draw-off nozzle. If not, the water would begin flashing to steam, as it experienced a pressure drop of 2 psi, flowing across the control valve. The evolved steam would then choke the water flow, reducing the pressure drop across the control valve until the pressure drop equaled the depth (or head) of water in the draw-off sump.

Control valve "B" is located 56 in below the draw-off nozzle. The pressure drop across the valve is still 2 psi, or 56 in of water. Again, we will neglect nozzle exit loss and friction loss. But in this case, the height of water in the draw-off sump may be zero. But why?

Referring to Fig. 11.7, the pressure head of water above the control valve "B" is an extra 2 psi, or 56 in of water. Thus, even if the control valve "B" loses 2 psi of pressure, the water will not flash to steam. And this is true even when the sump is almost empty.

Does this mean that control valves should be located well below the elevation of the draw-off nozzle? Yes!

Does this mean, that the amount of piping, fittings, gate valves, etc. should be minimized on a draw-off line, until the line drops 10 or 20 ft below the draw-off nozzle? Yes!

Does this mean any frictional losses, due to external piping, at the same elevation as the draw-off nozzle, have to be added to the nozzle exit loss, in determining the liquid level in the sump? Yes!

Overcoming Nozzle Exit Loss Limits

I once tried to increase the flow of jet fuel from a crude distillation column by opening the draw-off, flow-control valve. Opening the valve from 30 to 100 percent did not increase the flow of jet fuel at all. This is a sure sign of nozzle exit loss—or cavitation limits. To prove my point, I increased the level of liquid in the draw-off sump from 2 to 4 ft. Since the flow is proportional to velocity *and* the head is proportional

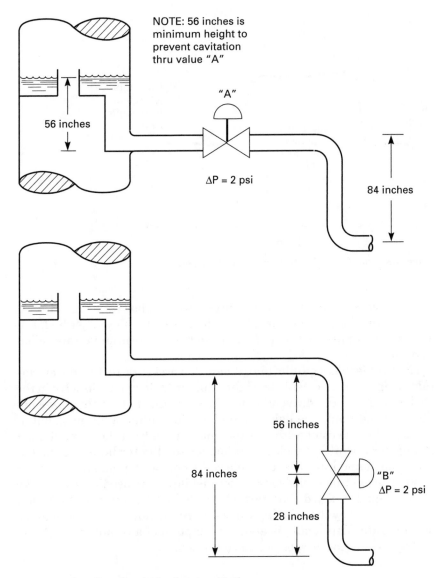

NOTE: 56 inches is minimum height to prevent cavitation thru value "A"

"A"

56 inches

ΔP = 2 psi

84 inches

56 inches

84 inches

"B"

ΔP = 2 psi

28 inches

Figure 11.7   Location of control valves is critical.

to (velocity)^2, the flow of jet fuel increased from 1000 to 1414 GPM. That is, the flow increased in proportion to the square root of the increase of liquid level in the sump. But the sump was only 4 ft high, and I needed to increase jet fuel flow even further. So, I raised the pressure in the crude column from 10 to 14 psig. This increased the pressure in the draw-off sump by 4 psi.

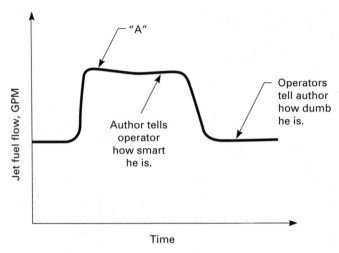

Figure 11.8 Equilibrium limits draw-off rates.

The results of this increase in pressure are shown in Fig. 11.8. At first, the increase in pressure greatly increased the flow of jet fuel. But after a few minutes, jet fuel flow slipped back to its original rate. What happened?

The problem is equilibrium. When the pressure in the tower was raised, lighter components from the vapor were forced to dissolve in the pool of liquid in the draw-off sump. This meant that, at the new equilibrium conditions, the lighter liquid would boil not at 10 psig, but at 14 psig. But the composition of the liquid already in the sump did not change for a while. It took some time for the lighter liquid, formed at the vapor-liquid interface, to displace the heavier liquid, already in the sump. During this period of time, jet fuel flow increased. But when the lighter liquid reached the draw-off nozzle, it began to vaporize. Flow was choked off by the evolving vapors, until the flow was reduced to the original rate. This again, is another symptom of a tower draw-off limited by nozzle exit loss, or nozzle cavitation.

Reference

1. Crane, *Flow of Fluids through Valves, Fittings, and Pipe,* Technical Paper no. 410, 25th printing, 1991, Crane Company, Joliet, Illinois.

12

Pumparounds and Tower Heat Flows

Closing the Tower Enthalpy Balance

There are two ways to remove heat from a distillation tower: top reflux and circulating reflux. In this chapter, we call a circulating-reflux stream a *pumparound.*

The vast majority of fractionators have top reflux. Cold liquid from the reflux drum is pumped onto the top tray of the tower. The cold liquid flashes to a hotter vapor. For example, let's say 1500 lb/h of liquid butane, at 100°F, flashes to 1500 lb/h of vapor at 260°F.

The specific heat of butane is 0.6 Btu/[(lb)(°F)]. The latent heat of butane is 130 Btu/lb (*latent heat* means the heat needed to change a pound of liquid into a pound of vapor at the same temperature). The heat removed by the top reflux is

$$(260°F - 100°F)(0.6)(1500) = 144,000 \text{ Btu/h} + (130)(1500)$$

$$= 195,000 \text{ Btu/h} = 339,000 \text{ Btu/h total heat removed}$$

The Pumparound

Pumparound heat removal

Figure 12.1 shows an alternate method, called *circulating reflux,* or *pumparound,* to remove heat from a tower. Hot liquid, at 500°F, is

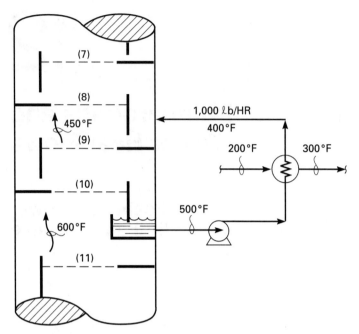

Figure 12.1 A pumparound or circulating reflux.

drawn from tray 10, which is called the *pumparound draw tray*. The liquid pumparound is cooled to 400°F. The cooled liquid is returned to the tower at a higher elevation onto tray 9. It appears from Fig. 12.1 that the cold 400°F pumparound return liquid is entering the down-comer from tray 8. This is often good design practice. Tray 9 is called the *pumparound return tray*.

The purpose of the pumparound is to cool and partially condense the upflowing vapors. The vapors to pumparound tray 10 are at 600°F. The vapors from the pumparound return tray 9 are at 450°F. There are two pumparound trays (9 and 10) in the column. This is the minimum num-ber used. A typical number of pumparound trays is two to five.

Let's calculate the heat removed in the pumparound circuit shown in Fig. 12.1. Assume that the specific heat of the pumparound liquid is 0.7 Btu/[(lb)(°F)]:

$$(500°F - 400°F) \times (0.7) \times 1000 \text{ lb/h} = 70{,}000 \text{ Btu/h}$$

I have not shown the flow of liquid on the cold side of the pumparound heat exchanger in Fig. 12.1, but we can calculate its flow. Let's assume that the specific heat of the cold-side liquid is 0.5 Btu/[(lb)(°F)]. Then the cold-side (or shell-side) flow is

$$\frac{(70{,}000\ \text{Btu/h})}{(300°F - 200°F) \times [0.5\ \text{Btu/[(lb)(°F)]}]} = 1400\ \text{lb/h}$$

What I have just illustrated is the most powerful tool in my bag of tricks—calculating a flow from a heat balance.

Purpose of a pumparound

Why do we wish to remove heat from the vapor flowing up through tray 10? The circulating pumparound is cooling the vapor flowing through tray 10, from 600 to 450°F, as it leaves tray 9. But for what purpose? Figure 12.2 shows the rest of the tower we have been discussing. Note that the tower-top reflux flow is controlling the tower-top temperature. If we were to reduce the pumparound circulation rate, less heat would be extracted from trays 9 and 10. More, and hotter, vapor would flow up the tower. The top reflux temperature control valve would open. The

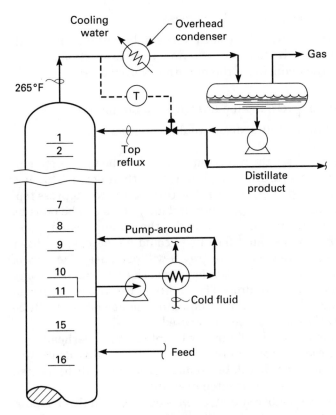

Figure 12.2 Two ways of removing heat from a tower.

top reflux rate would go up. The vaporization of reflux on the top tray would increase. The overhead condenser duty would increase.

The decrease in the heat duty of the pumparound heat exchanger would equal the increase in the heat duty of the overhead condenser. Thus, we say that the *heat balance* of the tower is preserved. Some of the heat that was being recovered to the cold fluid, shown in Fig. 12.2, is now lost to cooling water, in the overhead condenser. This shows the most important function of pumparounds: recovering heat to a process stream that would otherwise be lost to the cooling tower.

Let's say that the cooling-water outlet temperature from the condenser was 140°F. This is bad. The calcium carbonates in the cooling water will begin to deposit, as water-hardness deposits, inside the tubes. It is best to keep the cooling-water outlet temperature below 125°F, to retard such deposits. Increasing the pumparound heat removal will lower the cooling-water outlet temperature.

Another purpose of the pumparound is to suppress top-tray flooding. If tray 1 in Fig. 12.2 floods, the operator would observe the following:

- The tower-top temperature would increase.
- The distillate product would become increasingly contaminated with heavier components. If this were a refinery crude fractionator, we would say that the endpoint of the naphtha overhead product would increase.
- The pressure drop across the top few trays would increase.
- The liquid level in the reflux drum would increase.

If the operator increases the reflux rate, to reduce the tower-top temperature, the top temperature will go up rather than down. This is a positive indication of *top-tray flooding*. The correct way to suppress top-tray flooding is to increase the pumparound duty. This can be done by increasing the cold-fluid flow through the pumparound heat exchanger. Or, the pumparound flow itself could be increased. Either way, the flow of vapor flowing up to tray 8 will decrease. The flow of vapor through trays 1 to 7 will also decrease. The low vapor velocity will reduce the tray pressure drop. The ability of the vapor to entrain liquid will be reduced. The height of liquid in the downcomer will be reduced, and tray flooding will be suppressed.

Increasing pumparound heat duty will unload the overhead condenser. This will cool off the reflux drum. A colder reflux drum will absorb more gas into the distillate product. Less gas will be vented from the reflux drum, and this is often desirable.

Finally, is is best remembered that, as we said earlier in this chapter, heat recovered in the pumparound heat exchanger is often a valuable way to recover process heat. Heat not recovered in the pumparound exchanger is lost to cooling water in the overhead condenser.

Do pumparounds fractionate?

The process design engineer typically assumes that a pumparound is simply a way to extract heat from a tower. The engineer does not expect the trays used to exchange heat between the hot vapor and the cold liquid to also aid in fractionation. In practice, this is not what happens.

Let's refer back to Fig. 12.1. Note that the vapor temperature leaving tray 9 is 450°F. The temperature of the liquid leaving tray 10 is 500°F. This sort of temperature difference shows that fractionation is taking place across the pumparound trays. The temperature difference between

(Temperature of liquid leaving a lower tray)

− (temperature of vapor leaving a higher tray)

is a measure of the amount of fractionation. The bigger this temperature difference, the more the fractionation that is taking place across the trays. I have run plant tests which indicate that increasing pumparound circulation rates increases this temperature difference— up to a point. However, if the circulation rate is increased past this point, then the temperature difference is reduced. This idea is expressed best in Fig. 12.3. As the pumparound rate is increased, tray efficiency is improved. However, at some point, the pumparound liquid flow becomes too great. Probably, at this point the downcomers start to back up. Tray efficiency is impaired, because of this downcomer flooding. The temperature difference between the liquid leaving the

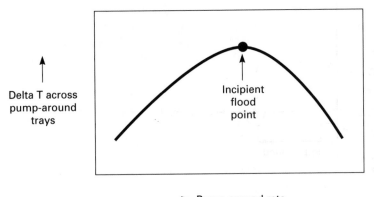

Figure 12.3 Pumparound trays do fractionate.

pumparound draw tray, minus the temperature of the vapor leaving the pumparound return tray, becomes smaller. This point is called the *incipient flood point,* for the pumparound trays.

Vapor Flow

How top reflux affects vapor flow

Let's assume that the vapor flow into a tower is constant. Figure 12.4 shows such a situation. Both the pounds per hour and temperature of the vapor flowing up to tray 9 is constant. This means that the heat flow into the tower is constant.

Now, let's increase the reflux rate. Certainly, the result will be

- The tower-top temperature will decrease.
- The gasoline overhead product flow will also decrease.

But what will happen to the flow of the vapor leaving tray 3, 4, or 5? I ask this question assuming that the pumparound heat duty is fixed. Will the pounds per hour of vapor flowing through trays 3, 4, or 5:

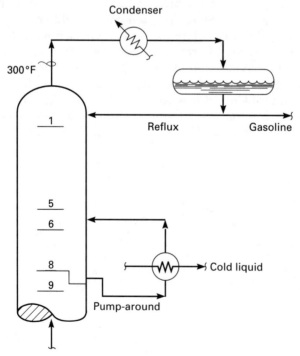

Figure 12.4 Effect of top reflux on vapor flow.

- Increase?
- Decrease?
- Remain the same?

The correct answer is that the weight flow of vapor will increase! Surprised? Most people are. Let me explain.

When we increase the reflux rate, the tower-top temperature drops—let's say from 300 to 240°F. Actually, the temperature of the vapor leaving all the trays in the tower will decrease. The effect is bigger on the top tray, and gradually gets smaller, as the extra reflux flows down the tower. If the top-tray temperature has dropped by 60°F, then the vapor temperature leaving tray 9 might drop by only 5°F. Let's assume that the extra reflux causes the temperature of the vapor from tray 4 to decrease by 40°F. We can say that the sensible-heat content of the vapor has decreased. *Sensible heat* is a measure of the heat content of a vapor, due to its temperature. If the specific heat of the vapor is 0.5 Btu/[(lb)(°F)], then the decrease in the sensible-heat content of the vapor, when it cools by 40°F, is 20 Btu/lb.

A small portion of this 20 Btu is picked up by the increased liquid flow leaving tray 4. The main portion of this heat is converted to *latent heat*. This means that some additional liquid on the tray turns into a vapor. But where does this extra liquid, which vaporizes on the tray, come from? It comes from the extra top reflux.

The vaporization of the extra reflux cools the tray. The extra vapor generated adds to the vapor flow from the tray. This increases the vapor flow from the tray. Even though the heat flow into the tower is constant, increasing the top reflux does increase the pounds of vapor flowing up the tower.[1]

Reflux affect on vapor molecular weight

Let's assume the following, for the tower shown in Fig. 12.4:

- Constant vapor flow to tray 9
- Tower-top reflux flow increased
- Pumparound duty constant

As a result of the increased reflux rate, the

- Tower-top temperature drops
- Gasoline flow drops

This is the same problem as I discussed previously. It will result in an increase in the mass flow of the vapor through trays 3, 4, and 5. But what will happen to the *molecular weight* of the vapor? Will it increase, decrease, or remain the same?

The answer is that the vapor's molecular weight will decrease, as the reflux rate is increased. But why?

The vapor leaving each tray is in equilibrium with the liquid. This means that the vapor leaving each tray is at its dew point and the liquid leaving each tray is at its bubble point. As the top reflux rate is increased, all the trays are cooled. The vapors leaving trays 3, 4, and 5 are cooled. As a vapor at its dew point cools, the heavier components in the vapor condense into a liquid. The remaining vapors have a lower molecular weight because they are lighter. But this is only half the story. Let us continue!

As the heavier components in the vapor condense into a liquid, they give off heat. This heat is called the *latent heat of condensation.* This latent heat is picked up by the liquid flowing across the tray. This liquid flow is called the *internal reflux.* This latent heat promotes extra vaporization of the internal reflux. Naturally, the lighter, lower-boiling-point components preferentially vaporize from the internal reflux. These lighter components have a relatively low molecular weight.

The uncondensed vapors flowing from the tray below, plus the newly vaporized vapors from the reflux, flow to the tray above. The combined molecular weight of vapors is thus reduced. As the molecular weight decreases, the volume of each pound of vapor increases.

As the molecular weight of the vapor decreases, the density of the vapor decreases. As the density of a vapor is reduced, each pound of vapor occupies more volume. It is true that cooler vapors do occupy less volume per pound than do warmer vapors of the same composition, as gases and vapors tend to contract on cooling and expand on heating. However, this is a small effect, compared to the increase in the volume of vapors, due to the decrease in the vapor's molecular weight.

Reflux affect on vapor volume

Now, let's summarize the affect on vapor flow due to increased reflux. Again, we are considering trays 3, 4, and 5 in Fig. 12.4:

1. Increasing the top reflux increases the pounds of vapor flow.

2. Increasing the top reflux increases the volume per pound of the vapor.

3. Increasing the tower-top reflux rate increases the rate (in ft³/s) of vapor flow through the trays, because of the combined, additive affect of factors 1 and 2.

How, then, does increasing the top-tray reflux rate affect the vapor and liquid loading on the trays below? Obviously, the liquid flow rate increases. But so does the vapor flow rate.

Does this mean that trays 3, 4, and 5 could flood, because the top reflux rate was increased? Yes!

Does this mean that we could flood these trays, without increasing the heat flow into the tower? Yes!

Does this mean that we could flood these trays without increasing the pounds of vapor flow into the tower, simply by raising the reflux rate? Absolutely correct!

If an increase in the tower-top reflux rate causes the top of the tower to flood, how should the operator respond? She should then increase the pumparound flow to reduce the pounds of vapor flow to tray 5, in Fig. 12.4. But suppose this causes the pumparound trays 6, 7, and 8 to flood, because of the extra liquid flow? She should increase the cold liquid flow through the pumparound heat exchanger. If this cannot be done, either, then the tower pressure can be increased. This will increase the density of the flowing vapors and shrink the volume of the vapors which the trays must handle.

Fractionation

Improving fractionation

Here is our problem. A refinery crude distillation tower is producing gasoline, truck diesel, and a gas oil. The diesel is contaminated with the gas oil. Also, the gas oil is contaminated with the lighter diesel. As shown in Fig. 12.5, the vapor flow into the tower is constant. Our job is to improve the *degree of fractionation* between diesel and gas oil. Our objective is to remove the relatively heavy gas oil from the diesel and to remove the lighter diesel from the gas oil.

We could reduce the amount of diesel product from the tower. That could wash the heavier gas oil out of the diesel. But it would also increase the amount of diesel in the gas oil. Increasing the heat removed in the pumparound would have a similar effect: less gas oil in diesel, but more diesel in gas oil.

How about decreasing the heat removed in the pumparound? This would seem to allow the lighter diesel to more easily vaporize out of the gas-oil product. But will this action result in increasing the contamination of diesel, with heavier gas oil? This answer is no—but why not?

Reducing the pumparound heat-removal duty increases the vapor flow from tray 8 in the column shown in Fig. 12.5. The extra pounds of

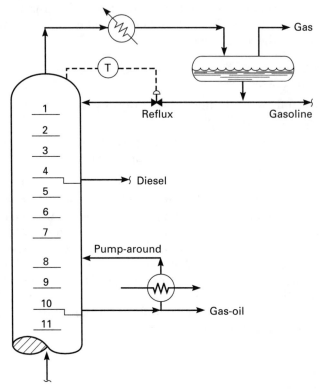

Figure 12.5 Decreasing pumparound improves fractionation.

vapor flow up the tower, and raise the tower-top temperature. The reflux control valve opens to cool the tower-top temperature back to its temperature set point. Then the liquid flow rates, from trays 1, 2, and onto tray 4, all increase. If the diesel draw-off rate is maintained constant, the liquid overflow rate onto trays 5, 6, and 7 will increase. This liquid flow is called the *internal reflux*. Trays 5, 6, and 7 are the trays that fractionate between diesel and gas oil. The more efficiently they work, the less the contamination of the adjacent products.

The way we increase the fractionation efficiency of trays is to make the trays work harder. The correct engineering way to say this is: "To improve the separation efficiency between a light and heavy product, the vapor flow rate through the trays is increased, and the internal reflux flowing across the trays is increased."

Again, this improvement in the degree of fractionation developed by trays 5, 6, and 7 is a result of reducing the amount of heat duty removed by the pumparound flowing across trays 8, 9, and 10.

Flooding the fractionation trays

Reducing the pumparound duty increases the tray loadings on trays 1 through 7. But in so doing, the trays operate closer to their incipient flood point. This is fine. The incipient flood point corresponds to the optimum tray performance. But if we cross over the incipient flood point, and trays 5, 6, and 7 actually start to flood, their fractionation efficiency will be adversely affected. Then, as we decrease the pumparound heat-removal duty, the mutual contamination of diesel and gas oil will increase.

From an operating standpoint, we can see when this flooding starts. As we decrease the pumparound duty, the temperature difference between the diesel- and gas-oil product draws should increase. When these two temperatures start to come together, we may assume that we have exceeded the incipient flood point, and that trays 5, 6, and 7 are beginning to flood.

Reference

1. For multicomponent mixtures, the assumption made when constructing a McCabe-Theile diagram of equal molal overflow does not apply.

13

Condensers and Tower Pressure Control

Hot-Vapor Bypass: Flooded Condenser Control

The total condensation of a vapor to a liquid is best illustrated by the condensation of steam to water. Figure 13.1 is a rather accurate reproduction of the radiator that heated my apartment in Brooklyn. Steam flowed from the boiler in the basement. The steam condensed inside the radiator, and flowed back into the boiler, through the condensate drain line. This is a form of thermosyphon circulation. The driving force for the circulation is the differential density between the water in the condensate drain line and the steam supply line to the radiator (see Chap. 5, discussion of thermosyphon reboilers).

The bigger the radiator, the more heat is provided to a room. The bigger the radiator, the faster the steam condenses to water inside the radiator. A larger radiator has more *heat-transfer surface area* exposed to the condensing steam. Unfortunately, the radiator shown in Fig. 13.1 is suffering from a common malfunction. Water-hardness deposits have partly plugged the condensate drain line. Calcium Carbonate is a typical water-hardness deposit.

Anyone who has lived in an apartment house in Brooklyn will have had to occasionally bang on the radiator to increase heat flow. Banging on the radiator often breaks loose the carbonate deposits in the condensate drain line. The steam condensate, which has backed up in the

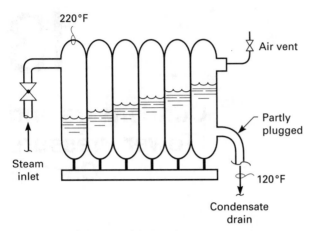

Figure 13.1   Radiator in a Brooklyn apartment house, circa 1946.

radiator, now empties. This exposes more of the interior surface area of the radiator to the condensing steam.

It rather seems that 40 percent of the surface area of the radiator in Fig. 13.1 is submerged under water. If the water is drained out, does this mean that the rate of steam condensation will increase by the same 40 percent. Answer—yes! Does this mean that the radiator heat transfer duty will increase by 40 percent? Answer—not quite.

Subcooling, Vapor Binding, and Condensation

Subcooling

Effect of subcooling.   When steam condenses at atmospheric pressure, it gives off 1000 Btu per pound of condensing steam. This is called the *latent heat of condensation of steam.*

When water cools off from 220 to 120°F, it gives off 100 Btu per pound of water. This heat represents the sensible-heat content of water between 220 and 120°F.

It takes less of the radiator's surface area to condense one pound of steam at 220°F than to cool off one pound of water from 220 to 120°F. And this is true, even though the condensation of steam generates 10 times as much heat as the cooling of hot water.

Does this mean that it is a lot easier to condense steam than to cool water? Yes. This also explains, then, why condensate backup reduces the rate of heat transfer and condensation.

Mechanics of subcooling.   As the condensed steam flow out of the radiator is restricted, the surface area of the radiator, available to cool the hot water, increases. Hence, the water temperature leaving the radiator decreases. To summarize, the effect of restricting the condensate flow from a radiator or condenser is to

- Build water level in the radiator.
- Reduce the rate of latent-heat transfer from the steam.
- Increase the rate of sensible-heat transfer from the condensate.
- Reduce the overall heat-transfer duty from the radiator.

Incidentally, the correct way to remove hardness deposit is by chemical cleaning. Violent banging on radiators is considered bad form in South Brooklyn.

Air lock

Vapor binding, or air lock, is another common cause of household radiator malfunction. Often, the vapor accumulating in the radiator is CO_2, rather than air. The CO_2 originates from the thermal decomposition of carbonates in the boiler. Regardless, air and CO_2 form a noncondensable vapor in the radiator. These noncondensables mix with the steam in the radiator. The noncondensables then reduce the concentration of the steam, by dilution. The diluted steam has a lower partial pressure than pure steam. The lower the partial pressure of the steam, the more difficult it is to condense. As the rate of condensation of the steam drops, so does the heat radiated by the radiator.

To restore the efficiency of a radiator, suffering from the accumulation on noncondensables inside its condensing coils, the noncondensable gases have to be removed. The air vent shown in Fig. 13.1 serves this purpose.

To summarize, the two most common malfunctions of a steam condenser (or radiator), are

- Condensate backup
- Noncondensable accumulation

And these two malfunctions are also the most common problems we encounter in the design and operation of shell-and-tube heat exchangers, used in total condensation service.

Condensation and condenser design

Condensation in shell-and-tube heat exchangers.   If what you have just read seems to be a repetition of the discussion on steam reboilers in

Chap. 8, your analysis is correct. A steam reboiler has the same problems, and works on the same principles as a process condenser. The only difference is that a steam reboiler's heat is removed by the shell-side process fluid, and a process condenser's heat is removed by cooling water.

Figure 13.2 is a sketch of a depropanizer overhead condenser. Let's make a few assumptions about this shell-and-tube condenser:

- The propane is totally condensed as it enters the reflux drum.
- There is no vapor vented in the reflux drum, but there is a vapor-liquid interface in the drum.
- The reflux drum is elevated by 20 ft above the top of the condenser.
- We are dealing with pure (100 percent) propane.

Well, if the liquid level in the reflux drum is located 20 ft above the condenser, does this mean that the liquid level in pipe feeding the condenser is also 20 ft above the condenser? Is this possible? No!

If the pipe to the condenser maintained a liquid level, then the shell side of the condenser would be full of propane. But, if the shell side of the condenser were really liquid full, the tubes would not contact the vapor. If the tubes do not contact the vapor, then the rate of condensa-

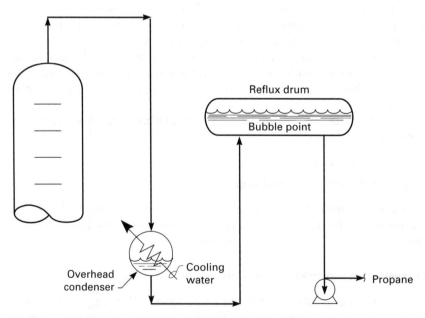

Figure 13.2 Total condensation below the reflux drum.

tion is zero. Perhaps a small amount of heat transfer would take place, as the liquid propane became subcooled. But none of the propane vapor would condense.

Therefore, the liquid level in the overhead condenser would have to be somewhere in the condenser's shell. But then, the liquid in the condenser would be below the reflux drum. How, then, does the liquid get from the lower elevation of the condenser to the higher elevation in the reflux drum? We will have to explain this hydraulic problem later. But for now, we can say that most reflux drums are elevated 20 or 30 ft above grade to provide *net positive suction head* (NPSH) for the reflux pump. Also, most shell-and-tube condensers are located at grade, for easier maintenance during unit turnarounds.

Subcooling in a shell-and-tube condensers. Figure 13.3 is the same propane condenser shown in Fig. 13.2. Let's assume that the pressure drop through the shell side is zero. Again, we are dealing with a pure component: propane. The inlet vapor is at its dew point. That means it is *saturated vapor*. Under these circumstances, the outlet liquid should be *saturated liquid*, or liquid at its bubble point. As the inlet dew-point temperature is 120°F, the outlet bubble-point temperature should be 120°F. But, as can be seen in Fig. 13.3, the outlet shell-side liquid temperature is 90°F, not 120°F. Why?

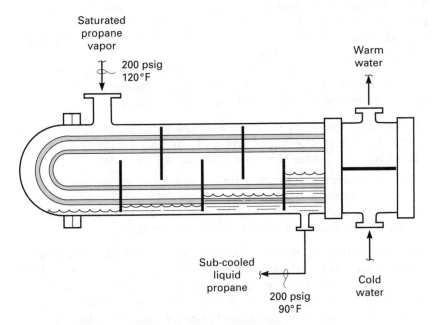

Figure 13.3 Condensate backup in a shell-tube heat exchanger.

The reason is condensate backup. The condensate backup causes subcooling; that is, the liquid is cooled below its bubble point, or saturated liquid temperature. Perhaps a rat has lodged in the condensate outlet pipe. The rat restricts condensate drainage from the shell side. To force its way past the dead rat, the propane backs up in the condenser. The cold tubes in the bottom of the shell are submerged in liquid propane. The liquid propane is cooled below its bubble-point temperature.

Note that the propane vapor is still condensing to propane liquid at 120°F. The condensed liquid is in intimate contact with the propane vapor, as it drips off the outside surface of the colder condenser tubes. The saturated propane vapor condenses directly to saturated propane liquid at 120°F. The saturated, or bubble-point, liquid then drips from the condensation zone of the condenser into the subcooling zone of the condenser. This is the zone where the tubes are submerged in liquid.

Try running your hand along the outside of such a condenser. Feel for the point on the surface of the shell where there is a noticeable drop in temperature. The upper part of the shell will be hot. The lower part of the shell will be cold. The transition point corresponds to the liquid level of condensate in the shell. The condensate level will always be higher toward the shell outlet nozzle. Again, this all applies only to condensers in total condensing service.

Effect of condensate backup. When the condensate level in an exchanger increases, the area of the condenser devoted to subcooling the condensate increases. But the area of the exchanger available for condensing decreases. That is bad!

When the area of the exchanger available for condensing is reduced, the ease of condensation is also decreased. Depending on circumstances, one of two unfavorable things will now happen:

1. If the supply pressure of the condensing vapor is fixed, the rate of condensation of the vapor will fall. In an apartment house in Brooklyn, this means that your bedroom will get cold.

2. If the condensing vapor flow rate is fixed, the condensation pressure, will increase. If you are operating a debutanizer in Texas City, this means that the tower's safety release valve will pop open, and release a cloud of butane over Lamar University.

Heat removal by condensation is easy. The overall heat-transfer coefficient U for condensation of pure, clean, vapors may be 400 to 1000 Btu per hour per ft^2 of heat exchanger surface area, per °F of temperature-driving force. The U value for subcooling stagnant liquid may be only 10 to 30. Condensate backup is the major cause of lost heat transfer for heat exchangers, in condensing service.

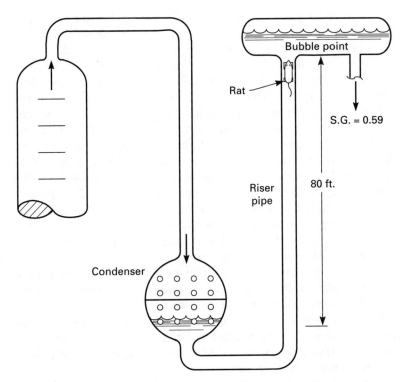

Figure 13.4 Elevation increase of reflux drum increases tower pressure.

Elevation increase promotes subcooling. Once upon a time many years ago, a tragic event occurred in Louisiana. A rat entered the condenser outlet pipe shown in Fig. 13.4. The condenser had been off line for cleaning. The rat, having crawled up the riser pipe to the reflux drum, got its head stuck in the drum's inlet nozzle. Your author, unaware of the rodent's predicament, put the exchanger back into service. The condensed butane now flowed across the rat. The rat died. Well, we all must come to that end eventually, although perhaps not quite that exact end. Such is the way of all flesh.

This rat is called a "20-lb rat." Not that the rat weighed 20 lb. The 20 lb refers to the pressure drop of 20 psig that the liquid encountered as it flowed across the rat's now-lifeless body. Before the introduction of this pressure restriction, the butane entering the reflux drum was at its bubble point. Our question is, will the introduction of the rat, at the inlet nozzle, cause the butane, as it enters the vessel, to flash?

The term "to flash" is used to denote partial vaporization of the butane. Before the rat became stuck, the liquid entering the reflux drum did not flash.

We can be sure that the butane liquid did not flash after the introduction of the rat because no vapor was vented from the reflux drum.

But, let's assume that the first microsecond after the introduction of the rat, that the liquid did vaporize. The vapor so generated, would be trapped in the reflux drum. The pressure in the drum would increase. Not by 20 psig, but just a little. The small increase in pressure in the reflux drum would push up the liquid level in the condenser. The surface area of the condenser, available to subcool the liquid, would increase. The liquid temperature would be reduced. As the subcooled liquid flowed across the dead rat, its pressure would drop. The liquid's pressure would fall, to exactly that pressure that corresponds to the vapor pressure of the butane, at the temperature in the reflux drum.

The backup of butane liquid in the condenser would continue until the butane, leaving the condenser, was cold enough so that it would not flash as it flowed across the rat—that is, until *equilibrium conditions* had been reestablished in the reflux drum.

As the butane liquid level in the condenser increased, the area of the exchanger exposed to the condensing vapors would decrease. Let's assume that the tower's reboiler duty was constant. The vapor flow rate to the condenser would then be constant. To condense the same flow rate of vapor, with a shrinking exchanger surface area, the pressure of condensation must increase. The tower pressure would also go up, as the condenser pressure rose.

Forget about the rat. I made that story up. The real story is that the riser pipe, connecting the condenser outlet to the reflux drum, was undersized. Nat Taylor, the project engineer, specified a 4-in pipe, when an 8-in pipe was needed. The pressure drop through the pipe was then 32 times higher than the intended ΔP of 0.60 psig (ΔP in a pipe varies to the fifth power of the pipe's diameter). The resulting riser pipe pressure drop was 20 psig. This *frictional loss* of 20 psig in the pipe had the exact same effect on the condenser—and on the tower's pressure—as the 20-lb rat.

Forget about Nat. Forget about the rat. Both stories are pure fiction. The truth is that the elevation of the reflux drum was 80 ft above the condenser. The specific gravity of the butane liquid was 0.59. This means that 80 ft of liquid exerted a head pressure of about 20 psig:

$$\frac{80 \text{ ft}/2.31 \text{ ft} \cdot H_2O}{psi} \times \frac{0.59 \text{ SG butane}}{1.00 \text{ SG } H_2O} = 20 \text{ psig}$$

where SG = specific gravity.

This *elevation head loss* of 20 psig had the same effect on the condenser—and on the tower's pressure—as did the rat or Nat's frictional loss.

Common design error.   Please refer back to Fig. 13.2. How can the liquid from the condenser rise to the higher elevation in the reflux drum, without being pumped? Simple! The pressure head of the liquid leaving the condenser is converted to elevation as the liquid flows up into the reflux drum. This works fine, as long as the liquid leaving the condenser is sufficiently subcooled. By "sufficiently subcooled," I mean that when the lower-pressure liquid flows into the reflux drum, it has to be cold enough that it does not flash.

The liquid leaving the condenser is subcooled. The liquid entering the reflux drum is saturated liquid at its bubble point. Of course, the temperature of the liquid is the same at both points. The subcooled liquid is "subcooled" in the sense that its pressure is above the bubble-point pressure, at the condenser outlet temperature. It is this extra pressure, above the bubble-point pressure, that may be converted to elevation.

Pressure Control

Tower pressure control

For total condensers, there are three general schemes for controlling distillation tower pressure:

- Throttling the cooling water flow to the condenser
- Flooding the condenser
- Hot-vapor bypass around the condenser

Regardless of the method elected, the principal concept of tower pressure control is the same. We control the pressure in the reflux drum by manipulating the temperature in the reflux drum. The tower pressure then floats on the reflux drum pressure. To lower the tower pressure, we must first cool the reflux drum. This cuts the vapor pressure of the liquid in the reflux drum.

The oldest, most direct method of pressure control is throttling on the cooling-water supply. This scheme is shown in Fig. 13.5. Closing the water valve to the tube side of the condenser increases the condenser outlet temperature. This makes the reflux drum hotter. The hotter liquid in the reflux drum creates a higher vapor pressure. The higher pressure in the reflux drum increases the pressure in the tower. The tower pressure is the pressure in the reflux drum, plus the pressure drop through the condenser.

Throttling on the cooling water works fine, as far as pressure control is concerned. But, if the water flow is restricted too much, the cooling-water outlet temperature may exceed 125 to 135°F. In this tempera-

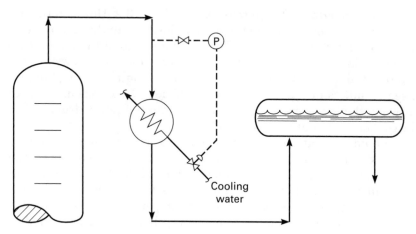

Figure 13.5 Tower pressure control using cooling-water throttling.

ture range, water-hardness deposits plate out inside the tubes. Then, the heat-transfer coefficient is permanently reduced by the fouling deposits.

Hot-vapor bypass

Hot-vapor bypass pressure control.   A more modern way of controlling a tower's pressure is shown in Fig. 13.6. This is the hot-vapor bypass method. When the control valve on the vapor bypass line opens, hot vapors flow directly into the reflux drum. These vapors are now bypassing the condenser. The hot vapors must condense in the reflux drum. This is because there are no vapors vented from the reflux drum. So, at equilibrium, the hot vapors must condense to a liquid on entering the reflux drum. They have no other place to go.

The latent heat of condensation of this vapor is absorbed by the liquid entering the reflux drum. The liquid that enters the reflux drum, comes from the condenser. The hot vapor mixes with the condenser outlet liquid, and is condensed by this cooler liquid.

Does this mean, then, that the condenser outlet temperature is lower than the reflux drum outlet temperature? If I were to place my hand on the reflux pump suction line, would it be warmer than the condenser outlet? Answer—yes!

But the liquid in the reflux drum is in equilibrium with a vapor space. This liquid is then at its bubble, or boiling, point. If the liquid draining from the condenser is colder than this bubble point liquid, then it must be subcooled. But how can a vapor condense directly into a subcooled liquid? Well, it cannot.

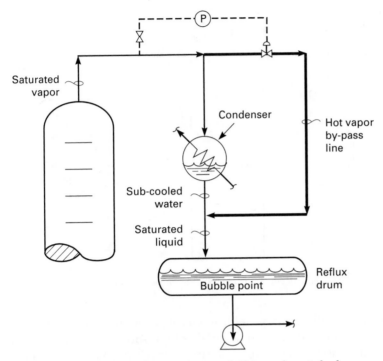

Figure 13.6 Hot-vapor bypass pressure control. Note condensate backup.

The tower overhead vapor, shown in Fig. 13.6, condenses to a liquid on the outside of the cold condenser tubes. The liquid drips off the tubes. These droplets of liquid are in close contact with the saturated vapor in the condenser shell. This means that the liquid is in equilibrium with the vapor. The condensed liquid is therefore, initially, at its bubble-point temperature. This liquid accumulates in the bottom of the condenser's shell. The submerged tubes then must subcool this liquid. Part of the surface area of the condenser is hence devoted to subcooling liquid, and part is devoted to condensing vapor.

But how does this work? How does the condenser know, without any advice from us, how much of its heat-exchanger surface area, which is supposed to be used to condense vapors, to divert to subcooling liquid?

Let's assume that the liquid draining from the condenser is not quite cold enough to absorb the entire latent heat of condensation of the vapors flowing through the hot-vapor bypass line. The vapors will then be only partially condensed. Vapor will start to accumulate in the reflux drum. This accumulation of vapor will increase the reflux drum pressure by a small amount. The higher drum pressure will back up the liquid level in the condenser by a few inches. The higher height of liq-

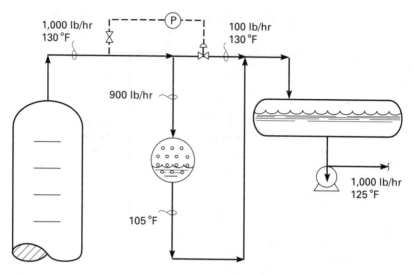

Figure 13.7   Opening pressure-control valve cools condenser outlet.

uid in the drum will submerge additional cold tubes with the condensed liquid. The cooler liquid will now be able to absorb more of the latent heat of condensation of the vapor passing through the hot-vapor bypass line. Eventually, a new equilibrium will be established.

Leaking hot-vapor bypass valve.   Let's assume that the hot-vapor bypass valve, shown in Fig. 13.7, is leaking. It is leaking 10 percent of the tower overhead flow. A good rule of thumb is then

- *Hydrocarbons.*   For each 20°F temperature difference between the cooler condenser outlet and the warmer reflux pump suction, 10 percent of the tower's overhead vapor flow is leaking through the hot-vapor bypass valve.

- *Aqueous systems.*   For each 20°F temperature difference between the condenser outlet and the reflux pump suction, 1 percent of the tower's vapor flow is leaking through the vapor bypass valve.

As the hot-vapor bypass valve opens, the condensate level in the shell side of the condenser increases to produce cooler, subcooled liquid. This reduces the surface area of the condenser exposed to the saturated vapor. To condense this vapor, with a smaller heat-transfer area, the pressure of condensation must increase. This, in turn, raises the tower pressure. This then is how opening the hot bypass pressure-control valve increases the tower pressure.

Incidentally, as I have shown in Figs. 13.6 and 13.7, the condenser may be located above or below the reflux drum. Both configurations require a subcooled liquid effluent from the condenser. But if the condenser is located below the reflux drum, additional subcooling to offset the elevation effect, described above, will be needed.

Flooded condenser pressure control

You should have concluded by now that hot-vapor bypass pressure control actually works by varying the surface area of the condenser exposed to the saturated vapor. But why do this indirectly? Why don't we simply and directly vary the liquid level in the condenser, as shown in Fig. 13.8?

The answer has something to do with women's clothes. Perfectly good, sensible designs go out of fashion. The perfectly good, sensible design, used to control a tower's pressure, *flooded condenser control,* shown in Fig. 13.8, has gone out of fashion. In a flooded condenser tower pressure-control strategy, the reflux drum is run full. Restricting the flow from the reflux pump increases the level in the condenser. This reduces the heat-transfer surface area available for condensation, and raises the tower pressure. Either the reflux or overhead product may be used to vary the liquid level in the condenser.

Once a liquid level reappears in the reflux drum, the condenser capacity has been exceeded. The level in the condenser will continue falling, until the drum empties, and the reflux pump begins to *cavitate.*

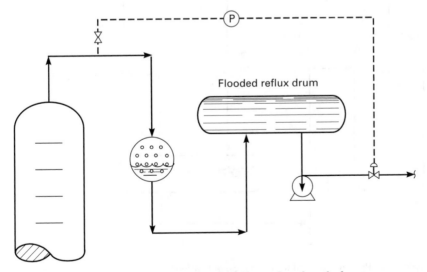

Figure 13.8 Flooded condenser pressure control: the preferred method.

In general, flooded condenser pressure control is the preferred method to control a tower's pressure. This is so because it is simpler and cheaper than hot-vapor bypass pressure control. Also, the potential problem of a leaking hot-vapor bypass control valve cannot occur. Many thousands of hot-vapor bypass designs have eventually been converted—at no cost—to flooded condenser pressure control.

The function of the reflux drum in a flooded condenser design is to

- Separate water from reflux when distilling hydrocarbons
- Give the operators time to respond, if they have exceeded the condenser's capacity
- Provide a place from which noncondensable vapors may be vented

Partial condensation

If we normally have a situation in which noncondensable vapors appear in the reflux drum, then there is only one pressure-control option available. This is to place the tower pressure-control valve on the vapor off-gas as shown in Fig. 13.9. If we normally have noncondensable vapors in the condenser effluent, then the following problems we have been discussing do not exist:

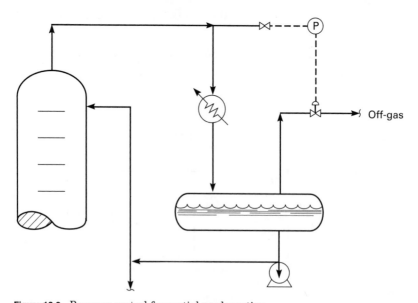

Figure 13.9 Pressure control for partial condensation.

- Condensate backup
- Subcooling of condenser effluent
- Fouling, due to low flow, of the cooling-water tubes

Sometimes we see tower pressure control based on feeding a small amount of inert or natural gas into the reflux drum. This is bad. The natural gas dissolves in the overhead liquid product and typically flashes out of the product storage tanks. The correct way to control tower pressure in the absence of noncondensable vapors is to employ flooded condenser pressure control. If, for some external reason, a variable level in the reflux drum is required, then the correct design for tower pressure control is a hot-vapor bypass.

14

Air Coolers

Fin-Fan Coolers

Air coolers are twice as expensive to purchase and install as water coolers. The great advantage of an air cooler is that it does not need cooling water. The difficult aspect of air cooling arises from the flow of air across the tubes.

Most air coolers are either induced-draft or forced-draft, as shown in Fig. 14.1. The more common arrangement being forced draft. The air is moved by rather large fans. The tubes are surrounded with foil-type

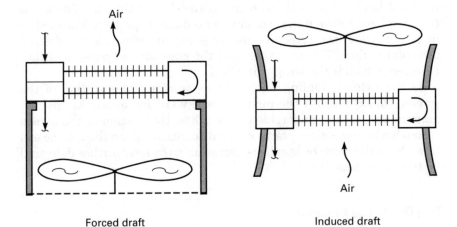

Forced draft   Induced draft

Figure 14.1 Two types of air coolers.

fins, typically one inch high. The surface area of the fins, as compared to the surface area of the tubes, is typically 12 to 1. That is why we call an air cooler an *extended-surface* heat exchanger.

The heat-transfer coefficient of an air cooler (Btu, per hour, per square foot of finned area, per degree Fahrenheit), is not particularly good. It might be 3 to 4, for cooling a viscous liquid, or 10 to 12 for condensing a clean vapor. The low heat-transfer coefficients are off-set by the large extended surface area.

Fin Fouling

In a forced-draft air cooler, cool air is blown through the underside of the *fin tube bundle*. In an induced-draft air cooler, cool air is drawn through the underside of the fin tubes. Either way, road dust, dead moths, catalyst fines, and greasy dirt accumulate along the lower row of tubes. As the tubes foul, they offer more resistance to the airflow. However, note that

- The total airflow discharged by the fan remains constant, regardless of the fin tube fouling.
- The fan discharge pressure remains constant, regardless of the fin tube fouling.
- The amperage electric load on the motor driving the fan remains constant, regardless of the fin tube fouling.

Figure 14.2 explains this apparent contradiction. As the underside of the fins becomes encrusted with dirt, an increasing amount of air is reflected back through the screen, located below the fan. The air is reflected back through the screen in a predictable pattern. The airflow in the center of the screen is always going up, which is the desired direction of flow. The airflow around the edge of the screen is always reversed, which is the wrong direction.

As the exterior fouling on the tubes worsens, the portion of the screen, through which the air flows backward increases. As the dirt accumulates on the underside of the tubes, the portion of the screen through which the air is drawn upward decreases. Even though the airflow blown through the bundle is decreasing, the total airflow delivered by the fan is constant.

Fan Discharge Pressure

Fan operation is indicated on a performance curve, as shown in Fig. 14.3. The head developed by the fan is equivalent to 5 or 10 in of water.

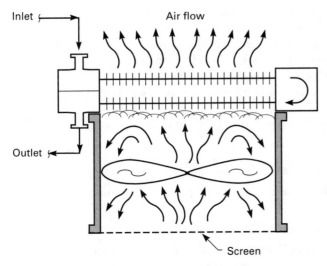

Figure 14.2 Airflow under partially plugged bundle.

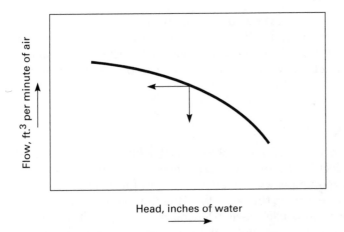

Figure 14.3 Fan performance curve.

As the fan airflow is pretty constant, the fan's head is also constant. Another way of stating this is to say that, as a tube bundle fouls, the resistance to airflow increases. This reduces the airflow through the bundle, but the pressure loss of the airflow through the tube bundle does not change.

If the head developed and the flow produced by a fan are both constant, then the power needed to run the fan must also be constant. Why? Because the power needed to spin a fan is proportional to the produced flow and the produced head.

To prove this to yourself, find the electric circuit breaker for a fan's motor. The amp (amperage) meter on the circuit breaker will have a black needle and a red needle. The black needle indicates the actual current, or amp load. The red needle is the amperage load that will trip the motor, as a result of *overamping*. Over time, as the tube bundle fouls and airflow through the bundle is restricted, the black needle never moves.

An induced-draft fan (see Fig. 14.1) is a different story. As the tube bundle fouls

- The air pressure to the fan drops.

- The air pressure from the fan is just atmospheric pressure. It remains constant.

- The water head (in inches) developed by the fan increases.

- The flow of air through the fan and the bundle decreases. This is consistent with Fig. 14.3.

- The amp load on the motor spinning the fan decreases.

Naturally, there is no reverse airflow on an induced-draft fan. That can occur only in a forced-draft fan. Reverse airflow can be observed with a forced-draft fan, by seeing which portions of the screen, shown in Fig. 14.2, will not allow a dollar bill to stick to the underside of the screen.

Effect of Reduced Airflow

Loss of airflow through a finned tube air cooler bundle is a universal problem. The effect is to reduce the exchanger's cooling efficiency. To restore cooling, you might wish to try the "Norm Lieberman method," which consists of reversing the polarity of the fan motor electric leads. The fan will now spin backward. Depending on the nature of the deposits, a portion of the accumulated dirt will be blown off the tubes—but all over the unit. Personnel observe this procedure from a "safe" distance.

A more socially acceptable option, is to water-wash the tubes. Most of the effective washing must be underneath the tubes. Washing from the top down is relatively ineffective. In many cases, detergent must be added to the washwater to remove greasy dirt. (*Caution:* Hot tubes may be thermally shocked by this washing, and pull out of the tube header box.)

The fan blades themselves may be adjusted to obtain more airflow. This is done by increasing the *fan blade pitch*. The pitch can usually be adjusted between 12° (for low airflow) to 24° (for high airflow). Any increase in airflow has to increase the amp load on the fan motor driver. The motor could then trip off.

Cooler weather always increases the airflow produced by a fan. This always increases the amp load on the fan's motor driver. To prevent the motor from tripping off, or simply to save electricity during the winter, you might reduce the fan blade pitch.

One factor that does not reduce airflow is crushed fins at the top of the tube bundle. Walking across a fin tube bundle will crush these fins. It looks bad, but does not appear to affect cooling efficiency.

Adjustments and Corrections to Improve Cooling

Adjusting fan speed

The revolutions per minute (RPM) (or rotational speed) of a fan can be increased by increasing the size of the *motor pulley,* which is the grooved wheel on the motor shaft. A small increase in the diameter of this pulley will greatly increase airflow through the cooling bundle. But, according to the *affinity* or *fan laws,* doubling the diameter of a pulley increases the driver amp load by 800 percent. That is, driver horsepower increases to the cube (third power) of the fan's speed.

But there is a bigger problem than motor overload when increasing a fan's speed. The blades themselves are rated only for a maximum centrifugal force. This force increases with increased fan RPM. At some maximum speed, the blades fly apart. Gentle reader, you can imagine how I became so smart on this subject.

Belt slipping used to be a major problem on air coolers. The resulting low RPM routinely reduced airflow. Modern air coolers have notched belts, which are far less subject to belt slippage. Regardless, a slipping belt will result in a reduced amp load on the fan's motor driver.

Use of water sprays on air coolers

Spraying water on fin-fan air coolers is generally not a good idea. It is really effective only in dry climates with low humidity. The evaporation of water by the dry air cools the surface of the fins; that is, the latent heat of vaporization of the water, robs sensible heat from the tubes.

Salts or other dissolved solids in the evaporating water will plate out on the exterior of the tubes. With time, a serious loss in heat-transfer efficiency results. Use of steam condensate can avoid this particular difficulty.

Water sprays should be used only as a stopgap measure, because of the swell they cause in the plant's effluent volume, and also their tendency to create a safety hazard in the vicinity of the cooler.

Designing for Efficiency

Tube-side construction

The mechanical construction of the tubes in an air cooler create some rather nasty problems. Figure 14.4 shows the exterior appearance at either end of an air cooler. The small black circles are threaded steel plugs. They are not connected to the ends of the tubes. Allow me to rotate the air-cooler header box, shown in Fig. 14.4, by 90°, and display a cross-sectional view in Fig. 14.5. Note that the plugs are not connected to individual tubes. Unscrewing a plug just gives one access to the end of a tube, for cleaning purposes.

Proper cleaning of an air-cooler tube requires removing two plugs. A large industrial air cooler may have 2000 tubes, or 4000 plugs. The labor involved to remove and reinstall all these plugs is formidable. Leaking plugs, due to cross-threading, is a common start-up problem. Hence, many air coolers are simply never cleaned.

The pass partition baffle shown in Fig. 14.5 makes this cooler a two-pass exchanger. These baffles are subject to failure, due to corrosion. More often, they break because of excessive tube-side pressure drop. The differential pressure across a two-pass, pass partition baffle equals the tube-side ΔP.

Once the pass partition baffle fails, the process fluid may bypass the finned tubes, and cooling efficiency is greatly reduced. This is bad. But worse yet, during a turnaround of the cooler, there is normally no way to inspect the pass partition baffle. There is no easy way to visually verify the mechanical integrity of this baffle. A few air coolers have removable inspection ports for this purpose; most do not.

Parallel air coolers

A large process plant air cooler may have 10, 20, 30, or more banks of air coolers, arranged in parallel. Figure 14.6 shows such an arrange-

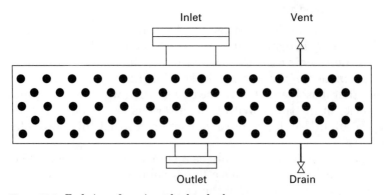

Figure 14.4 End view of an air-cooler header box.

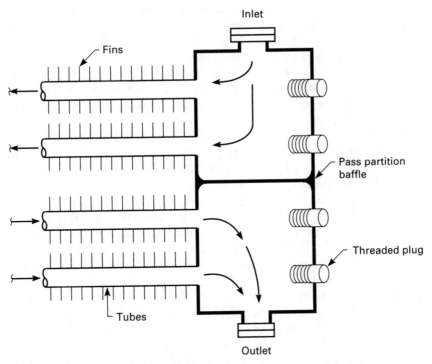

Figure 14.5 Cross section of an air-cooler header box.

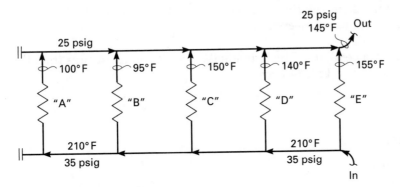

Figure 14.6 Air-cooler banks in parallel.

ment. Let's assume that the inlet header is oversized and has zero pressure drop. Let's also assume that the outlet header is oversized, and also has no ΔP. The pressure drop across the tube side, of all such air coolers arranged in parallel, is then identical.

If one of the air coolers begins to experience tube-side fouling, the fluid flow will be reduced. But the tube-side pressure drop will remain the same. The pressure drop across all five air-cooler bundles, shown in Fig. 14.6, is 10 psig.

Individual flows to parallel banks of air coolers are rarely—if ever—measured. Regardless, we can gauge the approximate relative flow to each bundle. This can be done by checking the outlet temperature of the bundles or banks.

Let's assume that the cooling airflow to all five banks are the same. Banks A and B in Fig. 14.6 have low outlet temperatures. Banks C, D, and E have must hotter outlets. Question: Which coolers are handling most of the heat-transfer duty? Is it A and B or C, D, and E?

The correct answer is C, D, and E. Most of the flow is passing through C, D, and E. Very little flow is passing through A and B. Look at the combined outlet temperature from all five coolers. It is 145°F. This indicates that most of the total flow is coming from C, D, and E—the banks with the higher outlet temperature. Very little of the flow is coming from A and B—the banks with the lower outlet temperature.

Why would the flow through A and B be so low? Apparently, their tubes must be partly plugged. Corrosion products, gums, and dust are common plugging agents. But when such exchangers foul, their relative tube-side ΔP, as compared to the other exchangers, remains constant. But their relative tube-side flow, as compared to the other parallel exchangers, decreases.

Air outlet temperature.   The individual air outlet temperatures from the coolers shown in Fig. 14.6 are

A: 105°F

B: 95°F

C: 170°F

D: 165°F

E: 180°F

These temperatures may be measured with a long-stem (24-in) portable temperature probe. Do not touch the tip of the probe to the fins when making a reading. Four readings per tube bundle section are adequate to obtain a good average.

The ambient temperature was 85°F. Then, the individual temperature rises for each air cooler would be

$$A = 105°F - 85°F = 20°F$$
$$B = 95°F - 85°F = 10°F$$
$$C = 170°F - 85°F = 85°F$$
$$D = 165°F - 85°F = 80°F$$
$$E = 180°F - 85°F = 95°F$$

$$\text{Total} \quad = \quad 290°F$$

If you are now willing to make the assumption that the airflow is the same through the five coolers, we could calculate the process side flow through each cooler. For example, percent flow through $A = 20°F/290°F$ = 7 percent. This calculation assumes that the percent of flow through the cooler is proportional to the temperature rise through the cooler, divided by the total temperature rise, through all five coolers.

It is not all that difficult to decide whether the airflow, through identical coolers, is similar. I just wave a handkerchief in the breeze at a few spots above the cooler.

Air-cooler condensers

In many process plants, the pump alleys are covered with forced-draft, air-cooled condensers. Dozens of coolers are arranged in parallel. I have seen services where 300 MM Btu/h of condensation duty was easily handled by aerial cooling. All these large systems had one problem in common. They all tended to have higher flows through cooling banks connected closest to the inlet and/or outlet headers. The higher relative flows were indicated by both higher air outlet and higher process outlet temperatures. A good example of this is shown in Fig. 14.6.

The mechanism that causes this often severe flow maldistribution is based on low-temperature dew-point corrosion, as explained below. Here is a rather common example, assuming that the main corrodants are chlorides, in a hydrogen sulfide–rich, condensing hydrocarbon vapor:

- A small amount of ammonia is injected into the overhead vapor line to control the pH of a downstream water draw.

- Vapor-phase ammonium chloride is formed.

- Tube bundles nearest the inlet header tend to see perhaps 1 or 2 percent more flow than the tube bundles farthest from the inlet header.

- Bundles seeing the lower flows have slightly lower outlet temperatures.

- Lower temperature favors the *sublimation* of ammonium chloride vapor to a white saltlike solid.

- This salt is very *hydroscopic,* meaning that it will absorb and condense water vapor from the flowing hydrocarbon vapor stream, at unexpectedly high temperatures.
- The resulting wet chloride salts are very corrosive, especially to carbon steel tubes.
- A ferric chloride corrosion product is formed.
- This metallic chloride salt then reacts with the abundant molecules of hydrogen sulfide to produce hydrochloric acid and iron sulfide.
- The hydrochloric acid may then continue to promote corrosion.
- The iron sulfide (or pyrophorric iron) accumulates as a blackish-gray deposit inside the tubes.
- This deposit further restricts vapor flow through the low-flow tubes.
- The reduced flow causes a lower tube outlet temperature.
- The lower tube outlet temperature promotes higher rates of salt sublimation from vapor to a corrosive fouling solid.

Meanwhile, the air-cooler bundles nearest the inlet header tend to see a greater and greater percentage of the total flow, as the cooler bundles foul and plug. They tend to stay hot and clean, and those bundles farthest from the inlet header tend to run cool and dirty.

It is a general principle of heat exchange that low flows tend to promote fouling and fouling promotes corrosion. The corroded, fouled heat-exchanger surface retards flow and creates a vicious cycle. We will see this problem again in shell-and-tube heat exchangers, as discussed in Chap. 21.

The best way to handle the nonsymmetrical flow problem described above is to make the pressure drops in both the inlet and outlet tube bundle headers very small, as compared to the bundle pressure drop itself. Many of my clients add additional tube bundles in parallel with existing air coolers. This helps at first, but they find that the long-term benefits are quite disappointing, because of high pressure drop in the new header lines.

15

Deaerators and Steam Systems

Generating Steam in Boilers and BFW Preparation

The one thing that almost all process and chemical plants have in common is steam. Steam is almost universally used to transfer energy over reasonably short distances (i.e., less than a mile), for the following reasons:

- Steam has a high latent heat of condensation. This permits flowing steam to transmit about eight times more energy than that transmitted by a pound of butane.
- Steam is nonexplosive, and hence reasonably safe to handle.
- Steam is generated from water, which, while not free, is often inexpensive to obtain.

Steam is used for several purposes in a modern process plant:

- To spin turbines
- To provide process heat, mainly to reboilers and preheaters
- As a stripping medium in fractionation towers
- To increase the velocity of heavier liquid streams, typically in furnace tubes

- As a *lift gas,* to lift heavier liquids or solid catalyst streams, to a higher elevation
- As a purge medium, to prevent pressure taps and other instrument connections and relief valves, from plugging

If energy, for either motive or heating purposes, must be transmitted over long distances (i.e., more than a few miles), it is best to use electricity. The energy loss (due to both friction and ambient-heat radiation) is orders of magnitude greater for steam, as compared to electrical power. Most electricity is produced from steam, by use of a steam-powered generator. Typically, it takes 3 Btu of steam to generate 1 Btu of electricity. That is why it is best to use the steam directly, at least for heating purposes, if the distance between the boiler house and the consumer is small.

Boiler Feedwater

Sources of boiler feedwater

Steam is generated from *boiler feedwater.* For example, raw water drawn from the Mississippi River has three objectionable contaminants:

- Sand and silt
- Dissolved solids (typically carbonates)
- Oxygen

The particulate matter (sand and silt) is removed by flowing upward through a gravel-and-sand filter bed. The coarser gravel is in the lower part of the bed. The filter beds are backflushed periodically with clear water to clean off the accumulated mud.

The dissolved solids, such as calcium carbonates, are removed by hot-lime softening or demineralization.[1] Demineralized water (also called *deionized water*) typically has essentially all anions and cations removed by ion-exchange resin. Demineralized water is preferable to hot-lime-softened water as boiler feedwater for several reasons.

For one thing, steam produced from hot-lime-softened water will have some amount of silicates. These silicates tend to deposit on the rotor blades of turbines, which use the motive steam as a source of energy. The silicate fouling of the turbine blades reduces the turbine's efficiency. But, more importantly, from an operator's point of view, the silicate deposits eventually break off of the blades. This unbalances the rotor. An unbalanced rotor is the fundamental cause of vibration. Vibrations lead to damage of the shaft bearings and seals. Eventually, vibrations will destroy the turbine's internal components.

Also, the hot-lime-softened water has variable amounts of carbonate contamination. When boiler feedwater is converted to steam, the carbonate deposits will break down into carbon dioxide and hardness deposits.

The hardness deposits coat the inside of the boiler's tubes, interfere with heat transfer, and overheat the tubes. The carbon dioxide, which is also generated from the dissolved solids, creates more serious corrosion problems in downstream heat exchangers. When the steam condenses, the carbon dioxide may remain trapped in the reboiler or preheater as a noncondensable gas. Actually, there is no such thing as a noncondensable gas. Even CO_2 is somewhat soluble in water. As the CO_2 dissolves in the condensed steam, it forms *carbonic acid,* a relatively weak acid (pH typically between 5 and 6). Strong acids will have pH values of 1 to 2. Pure water has a pH of seven. Carbonic acid is particularly corrosive to carbon steel heat-exchanger tubes.

Oxygen is a particularly reactive element. Actually, oxygen is a very potent gas and reacts aggressively with exposed metal surfaces to form oxides. Of all the corrosive substances encountered in a process plant, few exceed oxygen in reactivity with steel pipes. If a considerable amount of oxygen is left in boiler feedwater, interior corrosion of the boiler tubes will be rapid. The bulk of the oxygen, or dissolved air, is stripped out of the boiler feedwater in a *deaerator.*

In most properly operated process plants, most of the boiler feedwater should be recovered *condensate;* that is, when steam condenses to water, the condensate produced should be recovered and returned to the boiler house. Often, the recovered condensate will first pass through a deoiling clay bed, to remove any hydrocarbon contaminants. To calculate the percent of condensate recovered, we need to know two numbers:

- Total steam production (SP)
- Total raw water flow (RW)

The percent condensate recovery (CR) can then be calculated:

$$CR = 100\% - 100 \times \frac{RW}{SP}$$

Plants with CR >70 percent are usually doing a good job of recovering condensate. CR values of <30 percent represent poor performance. Poor condensate recovery is expensive for two reasons. One reason is the energy and chemicals needed to prepare the boiler feedwater. But in recent years, there is a second, more expensive cost associated with poor condensate recovery: effluent wastewater treatment.

I well remember one Louisiana refinery in which boilers and heat exchangers suffered from constant steam-side corrosion and fouling.

Investigation showed that the intake water-treatment plant operators would occasionally partially open the raw-water bypass around the entire treatment plant. They had a good reason, however, which they explained to me. "Norm, y'all have to admit, some kind of water is better than no water at all."

Deaerators

Figure 15.1 is a sketch of a typical boiler feedwater deaerator. The lower portion of the deaerator acts as a surge vessel for the boiler charge pump. The working part of the deaerator is the smaller vertical section on top of this surge vessel. The vertical section is essentially a small, four-tray, steam stripper tower (see Chap. 10). Sometimes these trays are called *mixing shelves,* but their function is the same as any distillation tower tray: to bring the vapor (steam) into intimate contact with the liquid (soft water). The objective of the little tower is to drive off the light component (air, dissolved in the soft water) with the heavier component (steam).

The deaerator shown in Fig. 15.1 is operating at 250°F. This particular temperature corresponds to the boiling point of water, at 15 psig. We can say, then, that the temperature of the deaerator controls its pressure. But, it is equally correct to say that the pressure of the deaerator sets its temperature.

The 120°F soft water is heated to 250°F by the 50-psig heating steam. In most deaerators, the vast majority of the heating steam is

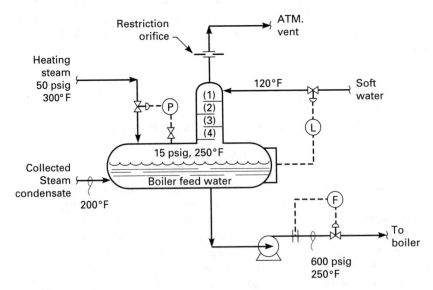

Figure 15.1 Deaerator is intended to remove O_2 from boiler feedwater.

condensed inside the deaerator. The heat of condensation of the steam is used to heat the soft water.

Usually, only a small amount of the heating steam—specifically, *stripping steam*—is vented through the restriction orifice, out the atmospheric vent. It is this small stripping steam flow that drives the air out of the soft water.

Note that the collected steam condensate does not pass through the little stripper section. As this stream is just condensed steam, it should be free of air. Hence, it does not need to be stripped.

A common deaerator malfunction. Big Jim was once a high-school teacher in New Orleans. I do not quite understand the connection, but many high-school teachers become really good process plant operators. Perhaps they find the work less hazardous. Big Jim became an operator at the utility complex for a major New Orleans area sugar refinery. Here is a rather neat problem that Jim and I solved together.

The problem was cavitation of the high-pressure boiler feedwater charge pump, shown in Fig. 15.1. Such pumps are often multistage. This means that there are three or four pump impellers on a relatively long shaft. The longer the pump shaft, the greater the damage caused by vibration to the pump support bearings and mechanical seal. Cavitation is certainly the leading cause of vibration in centrifugal pump operation (see Chap. 25). The cause of cavitation in this case was loss of liquid level in the deaerator surge or water storage vessel.

The symptoms of the problem Jim described to me were as follows:

1. An operator working at the main boiler would open the boiler's blow-down valve. Some lime-softened water, about 10 percent of the boiler's total feedwater must be drained to maintain the total dissolved-solids (TDS) content of the boiler's circulating water below a safe maximum.

2. This would draw water out of the deaerator vessel.

3. The level in the deaerator would fall, and the flow of 120°F soft water into the deaerator would increase.

4. Suddenly, the flow of softened water would increase exponentially. The level-control valve would rapidly open. Yet, the level in the deaerator would slowly continue to fall, until the pump lost suction and cavitated.

Jim and I then made the following field observations:

1. When the flow of softened water exceeded 80,000 lb/h, the heating steam pressure-control valve, shown in Fig. 15.1, would open to 100 percent.

2. A further increase in softened water flow drove the deaerator pressure down. This happened because the deaerator temperature also dropped.

3. Suddenly, the atmospheric vent pipe, which had been venting steam, started to blow out rather large quantities of water with the steam.

4. This stole water from the boiler feed surge portion of the deaerator.

5. As the deaerator water level fell, the 120°F softened-water flow increased.

6. The escalating flow of colder softened water cooled down the deaerator.

7. Since the water in the deaerator must always be at its boiling point, the reduction in the deaerator temperature resulted in a reduction in the deaerator pressure.

8. The reduced deaerator pressure coincided with an increase in water jetting out of the atmospheric vent.

What was happening?

Well, we have already discussed this phenomenon in Chap. 1, in the section on tray flooding. Water carryover from the top of tray 1, as shown in Fig. 15.1, is simply a clear signal that our little four-tray distillation tower is flooding. But what is causing the trays to flood?

Trays usually flood because of excessive vapor flow through them. The vapor flowing to tray 4 is essentially the heating steam flow. Certainly, most of this steam condenses on the trays, as the heating steam comes into contact with the 120°F softened water. It takes about 1 lb of 50-psig steam to heat 7 lb of water from 120 to 250°F. The uncondensed steam is then vented from the top of our tower through the atmospheric vent. But the full steam flow, from the pressure-control valve, does flow up through tray 4. Obviously, when this pressure-control valve is 100 percent open, the weight of vapor flow, to tray 4, is at its maximum.

But, dear reader, do not forget that when the steam inlet pressure-control valve is 100 percent open, any further increase of cool softened-water flow will suppress the deaerator's pressure. When the pressure of a vapor goes down, its volume goes up:

$$\text{Volume} \sim \frac{1}{\text{Pressure}}$$

And when the volume of a vapor flowing through a tray increases, so does its velocity. Any increase in vapor velocity through a tray results in higher tray pressure drop. And what is it that causes trays to flood? Why, it is high tray deck pressure drop.

To solve this problem, Jim recommended to his boss that the softened water be preheated by using it on one side of a heat exchanger to cool off a waste foul liquor stream which was flowing to the effluent treatment pond. This heated the softened water from 120 to 160°F. The 50-psig steam demand was reduced by 25 percent. This reduced steam flow prevented the trays from flooding.

Note that it is okay to heat softened or demineralized water to 160°F. You should not, however, do this with raw water. Calcium carbonate salts will plate out and foul the heat-exchanger tubes.

I rather like this problem. It illustrates how knowledge of one type of process equipment (a distillation tower) can help us understand problems in a piece of equipment that is superficially quite a bit different.

Boilers

A rather typical boiler is shown in Fig. 15.2. This is the sort of boiler one might rent to generate 50,000 lb/h of 150-psig steam. Incidentally, rented process equipment does not have to be returned to the owner if the plant files for bankruptcy. But that is another story.

The boiler shown in Fig. 15.2 relies on natural circulation. The density difference between the water plus steam in the downflow pipes and the riser pipes causes the water to circulate through the pipes. A typi-

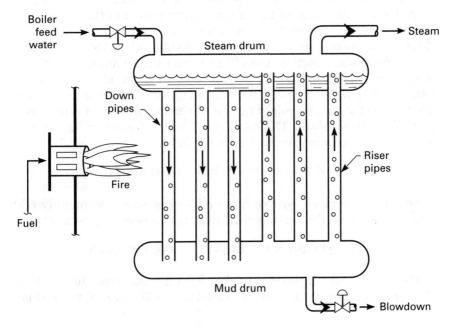

Figure 15.2 Natural-circulation boiler.

cal circulation rate results in 10 to 15 percent of the water flow being vaporized into steam. The produced steam flows from the steam drum into the steam header. The water level in the steam drum is maintained by the makeup boiler feedwater valve.

A certain amount of water is drained from the mud drum. The amount of this blowdown is determined by measuring the TDS in the steam drum. Typically, a target of several thousand parts per million (ppm) of TDS is specified. If the steam is generated at a low pressure (150 psig) for use in reboilers, a rather high TDS may be acceptable. If the steam is generated at a high pressure (1400 psig) for use in high-speed turbines, a much lower TDS target would be set.

It is really the quality of the treated boiler feedwater that sets the blowdown rate. Deionized or demineralized water might require a 1 to 2 percent blowdown rate. Hot-lime-softened water might require a 10 to 20 percent blowdown rate.

The amount of hardness deposits in steam is a function of entrainment of water, as well as the TDS of the boiler feedwater. A well-designed boiler, then, often is equipped with a mesh-type demister pad to remove entrainment from the produced steam.

Superheaters and economizers

Figure 15.2 is not a complete picture of a larger industrial-type boiler. Mainly, it does not show the superheat and economizer tubes. Figure 15.3 gives a better idea of the relative arrangement of the steam-generating tubes, superheat tubes, and economizer tubes.

Let's assume that we are generating 100,000 lb/h of 150-psig steam. In the economizer section, the effluent from the deaerator, at 250°F, would be heated to 350°F. As the specific heat of water is 1.0 Btu/[(lb)(°F)], we would need 100 Btu/lb of water. However, to produce 100,000 lb of steam, we might need a 10 percent blowdown rate to control TDS. This means 110,000 lb/h of boiler feedwater is needed. Therefore, the economizer heat duty would be

$$110{,}000 \text{ lb/h} \times 100 \text{ Btu/lb} = 11{,}000{,}000 \text{ Btu/h}$$

The latent heat of vaporization of 150-psig steam is roughly 880 Btu/lb. Therefore, the steam-generation section heat duty would be

$$100{,}000 \text{ lb/h} \times 880 \text{ Btu/lb} = 88{,}000{,}000 \text{ Btu/h}$$

The specific heat of steam is 0.5 Btu/[(lb)(°F)]. As shown in Fig. 15.3, the saturated steam will be superheated from 360 to 680°F. Or, we may

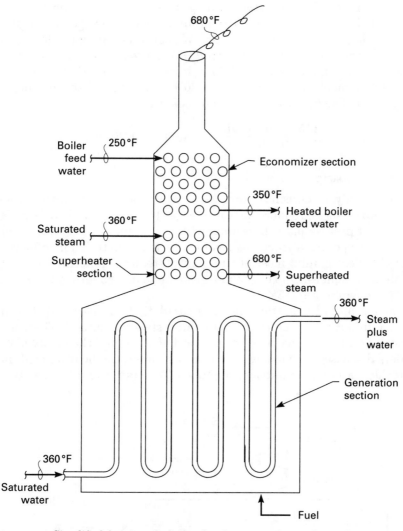

Figure 15.3 Simplified drawing of a boiler showing economizer and superheat sections.

say that the steam superheat will be 320°F. We would then need 160 Btu/lb. Hence, the superheater section heat duty would be

$$100{,}000 \text{ lb/h} \times 160 \text{ Btu/lb} = 16{,}000{,}000 \text{ Btu/h}$$

The total boiler heat duty would be the sum of the economizer, generation, and superheat sections, or 115,000,000 Btu/h.

I have noted that the stack, flue-gas temperature of the boiler is 680°F. Let's assume that we have around 3 percent oxygen in the flue gas (or about 15 percent excess air). We can then apply the following rule of thumb: *For every 300°F that the flue-gas temperature exceeds the ambient air temperature, the boiler efficiency drops by 10 percent.*

If our ambient air temperature is 80°F, then our boiler efficiency is 80 percent. The fuel consumed in our boiler is then

$$\frac{115 \times 10^6 \text{ Btu/h}}{0.80} = 144,000,000 \text{ Btu/h}$$

Waste-heat boilers

Probably only 20 percent of process plant operators or engineers ever work on the sort of boilers shown in Figs. 15.2 and 15.3. However, 90 percent of process plant personnel will, at one time or another, in all likelihood operate a *waste-heat boiler*. In many petroleum refineries and chemical plants, most of the plant's steam is generated in such waste-heat boilers. One of the most common such boilers is the kettle boiler, shown in Fig. 15.4.

The hot oil to the kettle boiler is a circulating pumparound stream, from a fluid catalytic cracker fractionator, slurry-oil circuit. There is a fundamental difference between this sort of boiler and the utility plant boilers discussed previously. In the kettle boiler, the heating medium is inside, rather than outside, the tubes. To obtain the full capacity of

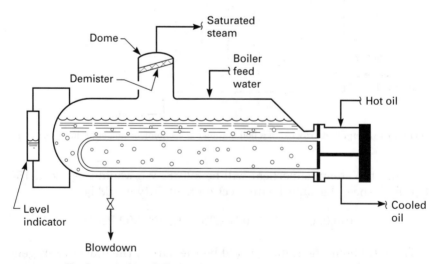

Figure 15.4 Kettle waste-heat boiler.

the kettle boiler, the uppermost tubes should be submerged in boiler feedwater. This requirement creates a few problems.

First, note the water level in the level indicator shown in Fig. 15.4. It looks quite a bit lower than the water level inside the kettle. Why? Well, because the water inside the kettle is boiling water. The water density has been reduced by the bubbles of steam. I have discussed this problem in great detail in Chap. 7, in the section on level control. Basically, the difficulty is that the water in the level indicator has a specific gravity of about 1.0. The boiling water in the kettle has a specific gravity of perhaps 0.5. If we have 4 ft of boiling water in the kettle, then, by pressure balance, we will have 2 ft of cool water in the level indicator.

To compensate for this problem as to the uncertainty of the water level in the kettle, the top tube of the tube bundle should be kept well below the top of the kettle. This also will help retard entrainment of boiler feedwater in the steam.

To further improve the steam quality, the experienced design engineer places a dome, as shown in Fig. 15.4, on top of the kettle. The dome will have a sloped demister mesh pad. A *demister* acts to coalesce small droplets of water into larger droplets. The larger droplets easily settle out of the flowing steam. The demister is sloped, to promote liquid drainage from the surface of the pad.

Of course, if the steam from the kettle will be used to reboil some nearby column, entrainment is not a major problem. If the steam is going into superheater tubes, however, serious entrainment will lead to solid deposits inside the superheater tubes, and this may cause these tubes to overheat and fail.

Convective Section Waste-Heat Steam Generation

Figure 15.5 shows another common form of steam generation from waste heat. Boiler feedwater is circulated, via a pump, through the convective tubes of a fired heater. Most of the heat released by the fuel (50 to 70 percent) is absorbed in the process coils. Perhaps 10 to 25 percent of the waste heat in the flue gas is absorbed in the steam-generation coils. It is good design and operating practice to keep the percent of water vaporized in the steam-generating coils, to well less than 50 percent. Higher rates of vaporization promote the laydown of salts inside the convective tubes. Low water circulation rates will lead to tube leaks.

It is probably best to consume steam generated in small waste-heat boilers, locally, in steam reboilers and preheaters. Turbine drivers are

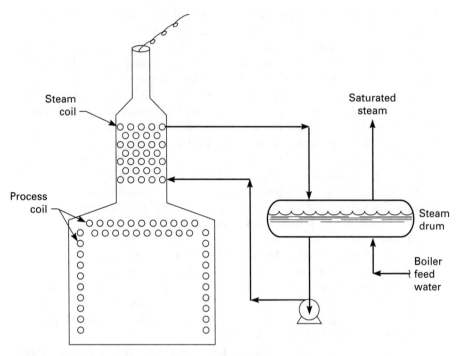

Figure 15.5   Convective section steam generation from waste heat.

best powered by boiler-house steam, where steam quality is more closely monitored and controlled.

Reference

1. *Betz Handbook of Industrial Water Conditioning,* 8th ed., copyright 1980, Betz Laboratories, Inc., Trevose, PA 19047 (USA).

Vacuum Systems: Steam Jet Ejectors

Steam Jet Ejectors

The converging-diverging steam jet is a startlingly complex device. Not only is the theory of operation rather weird, but the jets are subject to a wide range of odd, poorly understood, and never reported malfunctions. For all these reasons, I dearly love to retrofit and troubleshoot steam jet systems.

Steam jets have been around for a long time. They have just as ancient an origin as do steam-driven reciprocating pumps. They were used on early steam engines to pull a vacuum on the now-archaic barometric condenser. More recently, they were used to develop vacuums in such services as

- Surface condensers which condense the exhaust steam from steam turbines
- Petroleum refinery crude-residue vacuum towers
- Flash evaporators used to produce concentrated orange juice

Steam jets are also employed to recompress low-pressure steam to a higher-pressure steam. Jets are sometimes used to compress low-pressure hydrocarbon vapors with higher-pressure hydrocarbon gas (instead of steam). They are really wonderful and versatile machines.

Theory of Operation

The converging-diverging steam jet is rather like a two-stage compressor, but with no moving parts. A simplified drawing of such a steam jet is shown in Fig. 16.1. High-pressure motive steam enters through a steam nozzle. As the steam flows through this nozzle, its velocity greatly increases. But why? Where is the steam going to in such a hurry? Well, it is going to a condenser. The condenser will condense the steam at a low temperature and low pressure. It will condense the steam quickly. The steam accelerates toward the cold surface of the tubes in the condenser, where its large volume will disappear, as the steam turns to water.

The motive steam accelerates to such a great velocity that it can exceed the speed of sound (that is, exceed sonic velocity). This tremendous increase in velocity of the steam represents a tremendous increase in the kinetic energy of the steam. The source of this kinetic energy is the pressure of the steam.

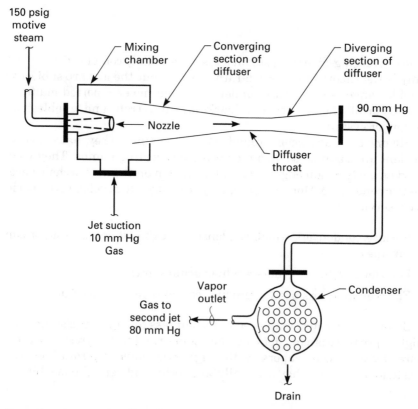

Figure 16.1 A converging-diverging jet.

The conversion of pressure to velocity is a rather common, everyday phenomenon. Remember Hurricane Opal, which struck the Florida panhandle in October 1995? It had peak winds of 145 MPH (mi/h). The pressure in the eye of the hurricane was reported at 27 in Hg. A portion of the kinetic energy of the wind in a hurricane is derived from the barometric pressure of the atmosphere. The lift which helps an airplane fly is also a result of the conversion of barometric pressure to velocity. Because of the shape of the wing, the air passes across the top of the wing at a higher speed than it does below the wing. The energy to accelerate the air flowing across the top of the wing comes from the barometric pressure of the atmosphere. Hence, the air pressure above the wing is reduced below the air pressure underneath the wing.

Similarly, as the high-velocity steam enters the mixing chamber shown in Fig. 16.1, it produces an extremely low pressure. The gas flows from the jet suction nozzle and into the low-pressure mixing chamber. It is not correct to say that the gas is entrained by the steam. The gas just flows into the mixing chamber, because there is a very low pressure in the mixing chamber.

The rest of the jet is used to boost the gas from the mixing chamber, up to the higher pressure in the condenser. This is done in two compression steps: converging and diverging.

Converging and Diverging Compression

Converging compression

When an airplane exceeds the speed of sound, we say that it breaks the sound barrier. In so doing, it generates a sonic wave or pressure wavefront. When steam and gas flow into the converging section of the jet diffuser shown in Fig. 16.1, the same thing happens. The gradually converging sides of the diffuser increase the velocity of the steam and gas, as the vapor enters the diffuser throat up to, and even above, the speed of sound. This creates a pressure wavefront, or *sonic boost*. This sonic boost, will multiply the pressure of the flowing steam and gas by a factor of perhaps 3 or 4.

Note something really important at this point. If, for any reason, the velocity of the steam and gas falls below the speed of sound in the diffuser throat, the sonic pressure boost would disappear.

Diverging compression

As the steam and gas leave the diffuser throat, the flow then enters the gradually diverging sides of the diffuser. The velocity of the steam and gas is reduced. The kinetic energy of the flowing stream is partially converted to pressure as the steam and gas slow down. This increase in pressure is called the *velocity boost,* which will multiply the pressure of the steam and gas by a factor of 2 or 3.

While smaller than the sonic boost, the velocity boost is more reliable. Even though the velocity in the diffuser throat in Fig. 16.1 falls well below the speed of sound, the increase in pressure in the diverging portion of the diffuser is only slightly reduced.

The overall pressure boost of a steam jet is obtained by multiplying the sonic boost effect times the velocity boost effect. The overall boost is called the jet's *compression ratio.*

Calculations, Performance Curves, and Other Measurements in Jet Systems

Vacuum measurement

We have discussed the American system inches of mercury (in Hg) in Chap. 6, "How Instruments Work." Of more immediate interest is Table 16.1. To do any sort of vacuum calculation, we need to convert to the absolute system, in millimeters of mercury (mm Hg). Unfortunately, we also need to correct measurements made with an American-type, (in Hg) vacuum gauge, for atmospheric pressure. You can interpolate between the two sets of data in Table 16.1, to correct for almost the entire range of typical atmospheric pressures.

TABLE 16.1   Vacuum Measurement Systems

Actual atmospheric pressure = 29.97 in Hg		Actual atmospheric pressure = 25.00 in Hg	
in Hg	mm Hg	in Hg	mm Hg
0	760	0	635
5.0	633	5.0	508
10.0	506	10.0	381
15.0	379	15.0	254
20.0	252	20.0	126
25.0	125	25.0	0
29.97	0		

We will need to use this table to calculate a jet's compression ratio, when we measure vacuum pressures with an American-type (in Hg) gauge.

Compression ratio

When considering the performance of a vacuum jet, we must first consider the jet's overall compression ratio. To calculate a jet's compression ratio:

Step 1. Measure the jet's suction pressure, and convert to millimeters of mercury, as shown in Table 16.1.

Step 2. Measure the jet's discharge pressure, and convert to millimeters of mercury.

Step 3. Divide the discharge by the suction pressure. This is the compression ratio.

It is not uncommon to find a proper jet developing an 8:1 ratio. More typically, jets will develop a 3:1 or 4:1 compression ratio. Any jet with less than a 2:1 compression ratio, has some sort of really serious problem.

Jet discharge pressure

The jet suction pressure is a function of the following factors:

- The overall jet compression ratio.
- The jet discharge pressure, as shown in Fig. 16.1.

The jet discharge pressure is controlled by the downstream condenser pressure. The minimum condenser pressure corresponds to the condensing pressure of steam at the condenser's vapor outlet temperature. For example, let's say that the condensing pressure of pure steam, at 120°F, is 80 mm Hg. If the condenser vapor outlet temperature is 120°F, then the lowest pressure we could expect to measure at the condenser vapor outlet would be 80 mm Hg.

Let's further assume that the pressure drop from the jet discharge, through the condenser discharge, is 10 mm Hg. Then, the jet discharge pressure would be 90 mm Hg. Let's also say that the sonic boost is equal to 3.60. The velocity boost is assumed to be equal to 2.5. The overall compression ratio is then

$$3.60 \times 2.5 = 9.0$$

The jet's suction pressure is then

$$(90 \text{ mm Hg})/9.0 = 10 \text{ mm Hg}$$

While I have seen steam jets develop compression ratios of 8:1 or 9:1, the majority of jets do not work nearly as well as that.

Multistage jet systems

A single jet, which discharges to the atmosphere, or to a condenser operating at atmospheric pressure, is called a "hogging" jet. Let's assume that atmospheric pressure is 29.97 in Hg or 760 mm Hg. Also, we will assume the jet is capable of an 8:1 compression ratio. Then, the jet's suction pressure would be

$$\frac{(760 \text{ mm Hg})}{8.0} = 95 \text{ mm Hg}$$

Referring to Table 16.1, this would mean that the jet inlet pressure was 26.2 in Hg, on an American-type vacuum pressure gauge. This is about the best we could expect with a single-stage jet.

Figure 16.2 shows a three-stage steam jet system. Let's first calculate the overall compression ratio for the combined effect of all three jets.

Note that atmospheric barometric pressure is 26 in Hg (this is a chemical plant in Denver—the Mile High City). The first-stage jet gas inlet pressure is 25 mm Hg. The third-stage jet discharge pressure is 3 psig. A good rule of thumb is

- 1 psi = 51 mm Hg

The actual barometric pressure can be converted to absolute mm Hg, using the following rule:

- 1 in Hg = 25.4 mm Hg

Therefore, the barometric pressure in Denver is

- 26 in Hg × 25.4 = 659 mm Hg

The pressure at the third-stage jet discharge is then

- 659 mm Hg + 3 × 51 mm Hg = 812 mm Hg

The overall compression ratio is

- 812 ÷ 25 = 32.5

But what is the average compression ratio for each of the three jets? Well, let's assume that the pressure drop for the first two condensers in Fig. 16.2 is zero. Then, let's remember that when compression stages work in series, their compression ratios are multiplied together, to calculate the overall compression ratio. Then the average compression ratio per jet is

$$(32.5)^{1/3} = 3.2$$

We take the cube root of the average compression ratio, because the three jets represent three compression stages.

Jet performance curves

I have rather implied, up to now, that a steam jet, depending on its mechanical condition, will develop a fixed compression ratio. This is not true. For one thing, the gas rate through the jet will influence its

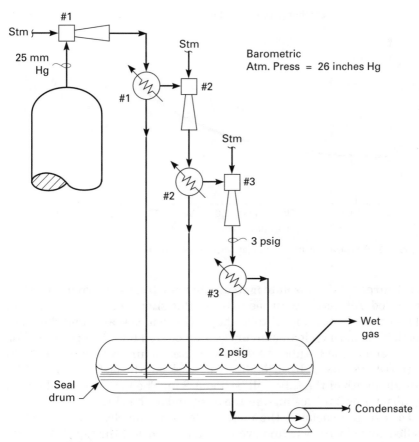

Figure 16.2 A three-stage jet system.

suction pressure. This is shown on the typical *jet performance curve,* in Fig. 16.3, which is drawn for a constant discharge pressure of 100 mm Hg. Let me make a critical comment about this curve. Variations in the jet's discharge pressure may have a surprising (i.e., nonlinear) effect on the jet's suction pressure. Sometimes, a small reduction in the discharge pressure will make a tremendous improvement in the suction pressure. The compression ratio might increase from 3:1 up to 7:1.

Sometimes, a very large reduction in the jet's discharge pressure, will not alter its suction pressure at all. The compression ratio might decrease from 7:1 to 3:1. It all depends on something called the *critical-flow* characteristics of the jet. More on this subject in a moment.

Measuring deep vacuums

For any vacuums better than 120 mm Hg (or 25 in Hg, at sea level), an ordinary vacuum pressure gauge will not be accurate enough for tech-

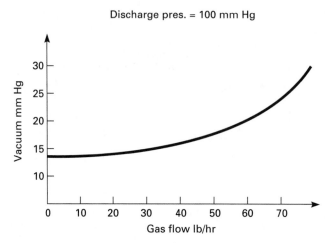

Discharge pres. = 100 mm Hg

Figure 16.3   A typical vacuum steam jet performance curve.

nical purposes. An absolute mercury manometer, as shown in Fig. 16.4, is needed. All that is required to make this simple device is a length of glass tubing bent into a U-tube shape. One end is sealed and the other end left open. Dry, clean mercury is then poured into the open end. The closed end of the U-tube is easily evacuated of air, by tipping the glass U-tube on its side. A little jiggling will work out the last air bubble. The overall length of the tube will be about 8 in. The mercury should wind up about 1 or 2 inches high in the open end of the U-tube.

To read the vacuum, the mercury level at the closed end must be pulled down by the vacuum even just a little, below the top of the tube. The difference in mercury levels between the closed and open ends of the U-tube, is the precise mm Hg vacuum.

Jet malfunctions

Big Spring is located in the scrub desert of western Texas. Cold autumn mornings are followed by warm afternoons. The local refinery's cooling-water temperature follows this ambient-temperature trend. The vacuum tower in this refinery also seems to keep track of the time of the day.

At 6:00 A.M., the primary steam jet is running quietly and pulling a vacuum of 12 mm Hg. At 7:30 A.M., the jet begins to make infrequent surging sounds. It rather sounds as though the jet is slipping every 15 or 20 seconds. As the morning coolness fades, the surging becomes more frequent. The vacuum also begins to slip from 12 to 14 mm Hg. Then, about 9:00 A.M., as the surges have become so frequent as to be almost continuous, the vacuum plunges to 23 mm Hg. By 10:00 A.M., surging has stopped and the vacuum at the jet suction has stabilized at

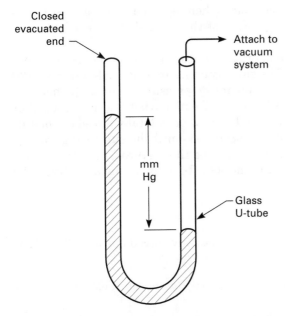

Figure 16.4 An absolute vacuum mercury manometer.

a poor 25 mm Hg. Many, if not most, operators of large vacuum jet systems have observed this problem—but what causes it?

Loss of sonic boost

Let's refer to Fig. 16.2. As the cooling water warms, the temperature of condenser 1 increases. This also increases the condensing pressure. This raises the discharge pressure of the primary jet (jet 1), as well as the pressure in the diffuser throat (see Fig. 16.1). Higher pressure will result in a smaller vapor volume. And a smaller vapor volume will cause a reduction of the velocity in the diffuser throat.

The lower velocity in the throat does not affect the jet's performance, as long as the velocity remains above the speed of sound. If the velocity in the throat falls below the speed of sound, we say that the jet has been forced out of critical flow. The sonic pressure boost is lost. As soon as the sonic boost is lost, the pressure in the vacuum tower suddenly increases. This partly suppresses vapor flow from the vacuum tower. The reduced vapor flow slightly unloads condenser 1 and jet 2 shown in Fig. 16.2. This briefly draws down the discharge pressure from jet 1. The pressure in the diffuser throat declines. The diffuser throat velocity increases back to, or above, sonic velocity. Critical flow is restored, and so is the sonic boost. The compression ratio of the jet is restored, and the vacuum tower pressure is pulled down. This sucks more vapor out of the vacuum tower, and increases the loads on condenser 1 and jet 2 (the secondary jet). The cycle is then repeated. Each of these cycles corresponds to the surging sound of the jet and the loss of its sonic boost.

As the cooling-water temperature rises, the sonic boost is lost more easily and more rapidly. The surging cycles increase in frequency to 30 or 40 per minute. The vacuum tower pressure becomes higher and higher. Finally, the surges become so frequent that they blend together and disappear. The primary jet has now been totally forced out of critical flow. The sonic boost has been lost until the sun in Big Spring, Texas sets and the desert cools. Surging then returns, until the critical flow in the jet is restored, and the sonic boost is regained at about 9:00 P.M.

Restoring critical flow

Steam jets, especially the primary jets, are forced out of critical flow, most commonly because

- Inadequate capacity of the primary condenser (i.e., condenser 1 shown in Fig. 16.5).

- Overloading, or poor performance of the second-stage jets.

The problem at Big Spring was rather typical. The two parallel, second-stage jets were not working as a team. Jet A in Fig. 16.5 was a real strong worker. Jet B was a loafer. It is rather like running two centrifugal pumps in parallel. Unless both pumps can develop about the same feet of head, the strong pump takes all the flow, and the weak

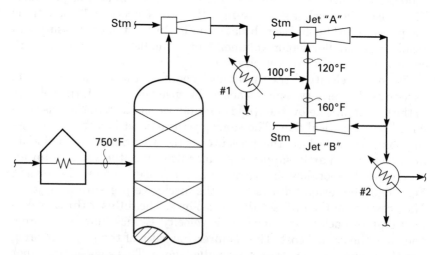

Figure 16.5 A malfunctioning secondary jet.

pump is damaged by internal recirculation. In the case of jets working in parallel, the strong jet takes all the gas flow from the upstream condenser. Furthermore, the strong jet sucks motive steam out of the mixing chamber (see Fig. 16.1). As you can see in Fig. 16.5, the suction temperature of jet A is 20°F hotter than the condenser 1 outlet temperature. This could happen only if the flow of vapor in the jet B suction line were backward. Blocking in jet B stopped the primary jet from surging until the more intense heat of the late afternoon.

Effect of gas rate

The most effective operating move to restore a jet to its critical-flow mode is to reduce gas flow. As I have described, this reduces the primary jet's (jet 1's) discharge pressure. Also, less of the energy of the motive steam is expended in accelerating the reduced gas flow. Hence, the steam enters the diffuser throat with greater kinetic energy. This also helps, along with the lower discharge pressure, in restoring critical flow and the jet's sonic boost.

To reduce the gas flow from the vacuum tower, shown in Fig. 16.5, I cut the heater outlet temperature from 750 to 742°F. This reduced thermal cracking in the vacuum heater and the consequent production of cracked gas. The pressure in the vacuum tower dropped from 21 to 12 mm Hg, and production of valuable heavy gas oil from the vacuum tower-bottom residue increased by 20 percent. I can still recall the warm afternoon sunshine on my face, as I signaled to the control-room operator to drop off that final degree of heater outlet temperature. I can still hear the last surge dying away, as the primary jet recovered from its long illness. And as the jet began its steady, full-throated roar, I knew it was running in its proper critical-flow mode.

Reducing primary-jet discharge pressure

Let's say that a jet is already in its critical-flow mode. It is already benefiting from both the sonic boost and the velocity boost. What, then, will be the effect of a reduction in the jet's discharge pressure on the jet's suction pressure? Answer—not very much. If a reduction in discharge pressure is made on a jet, which is *not* working in its critical mode, there will always be some benefit.

But if the jet is already in critical flow, reducing the pressure downstream of the diffuser throat cannot significantly raise the flow of gas into the diffuser throat. I know. I've tried. Twice I have added a third-stage jet to an existing two-stage jet system. The discharge pressure from the second stage jet dropped by 500 mm Hg. The discharge pressure of the first-stage jet dropped by 160 mm Hg. The suction pressure to the first-stage jet dropped by perhaps 2 mm Hg.

Condensate backup

The jet in Fig. 16.6 has a compression ratio of (180 mm Hg)/(150 mm Hg) = 1.20. This extremely low compression ratio does not indicate any sort of jet malfunction. The high jet suction pressure is caused by the 140°F precondenser outlet temperature. The vapor pressure of water at 140°F is 150 mm Hg. There is a large amount of process steam flowing into the precondenser. The lowest possible pressure that the precondenser can operate at, and still condense the process steam, is 150 mm Hg. As the jet sucks harder, it just pulls a few more pounds of water out of the precondenser, without altering the precondenser's pressure.

The problem with the precondenser is condensate backup. Something, perhaps a partially plugged drain line, is restricting condensate flow. As the condensate backs up, it reduces the surface area of the condenser, exposed to the condensing process steam. This makes it more difficult for the process steam to condense. The condensate backup also subcools the condensate. The net result is that the precondenser vapor outlet temperature goes up and the precondenser liquid outlet temperature goes down.

Recently, on a job in Arkansas City, I was able to force a jet to surge and lose its sonic boost, simply by raising the condensate level in its downstream condenser by just 6 in. Lowering the level drew down the jet's discharge pressure by a few millimeters of mercury and restored it to critical flow.

Jet freeze-up

There is another type of jet surging, which is caused by the motive steam turning to ice. How is this possible? Certainly, steam cannot turn to ice inside the jet? But it can and does.

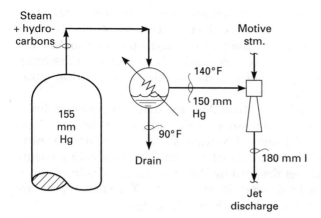

Figure 16.6 Condensate backup limits vacuum.

The jet system was in Mobile, Alabama. The symptoms of the problem were

- Extremely poor performance of the primary jet.
- The poor performance was constant, regardless of the cooling-water temperature.
- The jet would roar along in a normal fashion, and then go extraordinarily quiet for 10 to 15 seconds.
- The outside of the mixing chamber would chill to exactly 32°F during those times when the jet was quiet. This happened even though it was a bright, warm, sunny day in Mobile.

Incidentally, the way to measure surface temperatures in the field is with an infrared handheld thermometer. The response time of these thermometers is infinitely faster than in old-style, contact thermometers. You can buy one from any good instrument catalog[1] for a few hundred dollars.

Naturally, if the jet freezes, steam flow will stop. The jet will be quiet, and its compression ratio will be nil. But what causes the steam to turn to ice? Well, a number of factors extract heat from the steam:

- When any vapor expands, due to a pressure reduction (other than H_2 and CO_2), it cools off. This is called a *Joule–Thompson expansion*. The reduction in temperature of the steam is called a *reduction in sensible-heat content*. The sensible heat of the steam is converted to latent heat of condensation. Does this mean that the latent heat of condensation of 10-psig steam is much higher than that of 450 psig steam? Let's see:
  - Latent heat of condensation of saturated 10 psig steam = 980 Btu/lb.
  - Latent heat of condensation of saturated 450 psig steam = 780 Btu/lb
- When the velocity of a vapor increases, some of the increase in kinetic energy is extracted from the sensible heat of the vapor.
- The steam also gives up some of its energy to provide increased momentum to the gas flowing into the jet suction nozzle (see Fig. 16.1).

If the motive steam were dry, these factors would reduce the 150-psig motive-steam saturated temperature from 350 to about 100°F. But the motive steam in Mobile was not dry. It had partly condensed in the steam supply line to the jet. If steam is wet and contains liquid water, the water will flash to steam if the steam pressure is suddenly reduced to a vacuum. But the heat of vaporization comes from the sensible-heat content of the steam. If the steam contains 10 percent moisture, it will chill by 180°F on flashing. This implies that we could have the wet, motive-steam temperature dropping to −80°F as it enters the jet's mixing chamber. But, of course, the steam will turn to ice when its temperature drops to 32°F. The ice blocks the flow of steam. As the steam velocity slows, the jet warms and melts the frozen steam, and the steam flow is restored.

Wet steam

The problem in Mobile was resolved by installing a small steam filter on the steam line to the jet. This filter extracts moisture from the steam, and blows it out through a steam trap.

But wet steam is bad for a jet, even when it does not cause the jet to freeze. Mainly, wet steam causes erosion of the steam inlet nozzle. Erosion of this nozzle is the main reason why jets undergo mechanical deterioration. As the nozzle erodes, it allows more steam to pass through into the diffuser. The diameter of the diffuser is designed to operate with a certain steam flow. If that design steam flow is exceeded, the diffuser operation suffers. Also, the downstream condenser pressure will also increase.

An eroded steam nozzle shows no obvious sign of damage. The erosion is quite uniform and the nozzle interior is smooth. The inner diameter of the jet must be checked carefully with a micrometer. Growth in diameter of just 5 to 10 percent is significant. The nozzle is intended to be replaced periodically, much like the wear ring on a centrifugal pump.

Don't know why, but it happens

I have learned a lot about how process equipment actually works by investigating comments such as "It may not make any sense to you, but that is what happens here." Process equipment always conforms to the principles of science, but we have to know which principle to apply.

Figure 16.7 shows an old vacuum tower in Aruba. The chief operator on this unit made the following statements:

- "The colder the vapor outlet temperature from the precondenser, the better the vacuum he could pull, because of reduced vapor flow to the jet." Agreed!

- "Increasing cooling-water flow to the precondenser decreases the vapor outlet temperature." Agreed!

- "Closing the cooling water outlet valve A, about three-fourths of the gate valve steam travel, increases cooling-water flow through the precondenser." Nonsense!

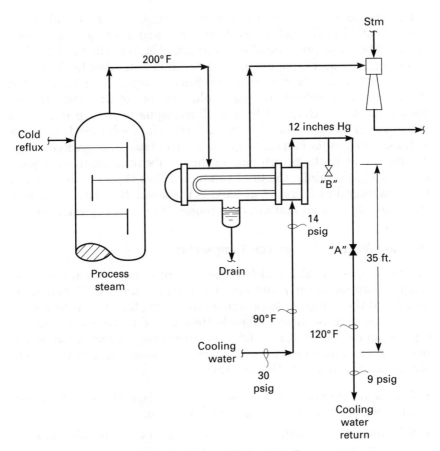

Figure 16.7 Air evolving from cooling water reduces water flow.

Really, dear reader, how can closing a valve in a pipeline increase flow in that pipeline? It cannot, and it will not, and it did not. Yet, on the other hand, it is a really bad idea to disregard field observations made by experienced plant operators. So, let's take a closer look at Fig. 16.7.

First, I tried opening valve A. Just as the chief operator said, the cooling-water outlet temperature increased, proving that water flow was reduced. Next, I checked the pressure at bleeder B; it was 12 in Hg. The pressure was so low at this point because of

- The 20-psi ΔP of the cooling water as it flowed through the tube bundle.
- The 35 ft of elevation (about 15 psi of head pressure) that the water had to gain to climb to valve B. Of course, this 15 psi of head loss was regained when the water flowed back down to the cooling-water return header.

The first idea I had was that the 120°F water would partially flash to steam at 12 in Hg and the evolved vapor would restrict water flow. Wrong! The vapor pressure of water at 120°F is 26 in Hg, not 12 in Hg.

But cooling water is not just pure water. It is water that has been saturated with air in the cooling tower. Sure enough, when I calculated the amount of dissolved air that would flash out of water (which had been saturated with air at 90°F and atmospheric pressure), I found that a very large amount of air could flash out of solution. It was just as the chief operator had said. Opening valve A too much, evolved large volumes of air in the return pass of the U-tube precondensers, shown in Fig. 16.7. Certainly, the total volumetric flow through the outlet of the condenser does increase, as valve A is opened. But the incremental flow is all air, and all that air, does choke off the cooling-water flow.

Optimum Vacuum Tower-Top Temperature

The chief operator also insisted that lowering the vacuum tower-top temperature too much would hurt the vacuum. But why? There is no doubt that the colder the tower-top temperature, the less the heat-duty load for the precondenser to absorb. Hence, cooling the vacuum tower-top temperature should, and did, reduce the precondenser vapor outlet temperature. This should have reduced the vapor load to the downstream jet. But it didn't. Here is why:

- The vapor components distilled overhead from the vacuum tower consisted of steam, cracked gas, and naphtha.
- The steam and naphtha vapors would pretty much totally condense in the precondenser, shown in Fig. 16.7.
- Some of the cracked gas would dissolve in the condensed naphtha. Most of the cracked gas would flow on to the jet.
- Increasing the tower-top temperature would distill over more pounds of naphtha.
- The extra condensed naphtha would dissolve more cracked gas.
- The reduced flow of cracked gas to the jet would unload the jet and permit it to develop a larger compression ratio.

Of course, if the vacuum tower-top temperature became too high, the increase in the precondenser vapor outlet temperature would increase the vapor pressure of water. This factor would then limit the minimum pressure in the precondenser.

In another case, a large volume of NH_3 was accidentally injected in the inlet to the condenser. The vacuum instantly improved. Why? Well, the NH_3 reacted with the H_2S in the cracked gas to form NH_4HS

(ammonium sulfide). This salt is very soluble in water; H_2S is not. The H_2S was effectively extracted from the cracked gas, and the downstream jet was thus unloaded and sucked harder.

Jets have been partially replaced by liquid ring-seal pumps. These are really positive-displacement compressors. The gas is squeezed between the vanes of the compressor's rotor and a pool of liquid in the compressor's case. Liquid ring-seal pumps are not interesting. They have no character. They are not as complex as steam jets. Anyway, I will discuss positive displacement compressors in later chapters.

Reference

1. *Davis Instruments* (catalog), vol. 60, copyright 1995, Davis Instruments (printed in USA, Davis Instruments), 4701 Mount Hope Drive, Baltimore, MD 21215.

17

Steam Turbines

Use of Horsepower Valves and Correct Speed Control

A *steam turbine* is a machine with an ancient genealogy. It is a direct descendant of the overshot water wheel, used to kick off the industrial revolution in England, and the windmill still used in Portugal. Turbines are widely used in process plants to drive everything from 2-hp pumps to 20,000-hp centrifugal compressors. They are versatile machines, in that they are intrinsically variable-speed devices. Electric motors are intrinsically fixed-speed machines. It is true that there are a variety of ways to convert AC (alternating-current) motors to variable speed, but they are all expensive and complex.

There are two general types of steam turbines: extraction and condensing. The most common turbine with which the process operator comes into contact is used to spare an electric-motor-driven centrifugal pump. The three-phase, AC motors used in the United States and South America are either 1800 or 3600 rpm. The motors used in Europe are 1500 or 3000 rpm. Small steam turbines (20 to 500 hp) used to drive centrifugal pumps, are rated for the same speed as the electric motors that are used in that particular service.

Principle of Operation and Calculations

Did you ever turn a bicycle over on its handle bars, and squirt water from a garden hose at its front wheel? What causes the wheel to spin with such great speed? What is the property of water, striking the spokes, that causes the wheel to spin? Is it the pressure of the water hitting the spokes that spins the wheel? Certainly not! While the water pressure in the hose might be 40 psig, as the water discharges from the hose nozzle, its pressure surely falls to the atmospheric pressure in our garden. But what happens to the pressure of the water? It is converted to velocity!

It is the velocity of the water, then, that causes the wheel to spin. The velocity of the water is transferred to the spokes of the wheel.

What force of nature causes a windmill to spin? Answer: the wind (or the velocity of the air). What causes a steam turbine to spin? Answer: the velocity of the steam hitting the turbine wheel.

A simple steam turbine

Figure 17.1 is a conceptual drawing of a simple topping steam turbine. An actual turbine does not look anything like this sketch. The 400-psig

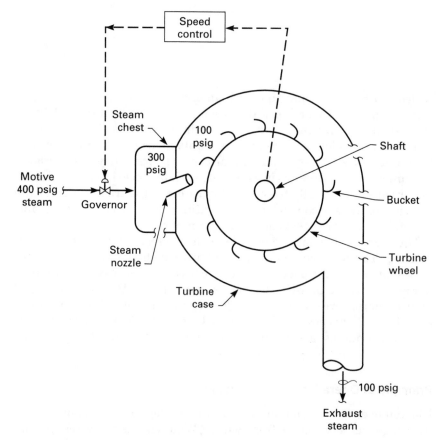

Figure 17.1 Conceptual sketch of a steam turbine.

motive steam enters through a *governor speed-control* valve. This valve, sometimes called the *Woodward governor,* controls the flow of steam into the *steam chest.* If the turbine is running below its set speed, the governor valve opens.

It is really just like cruise control on your car. You select the speed you want to drive at. If the car is going slower than the speed you selected, or set speed, the gas pedal is depressed, to bring the car up to your selected set speed. The governor speed-control valve is just like the gas pedal.

The pressure drop across the governor control valve in Fig. 17.1 is 100 psi. This sort of pressure drop is called an *isoenthalpic expansion*— a term used in thermodynamics meaning that the energy of the steam has been preserved during its reduction in pressure from 400 to 300 psig. It is also an *irreversible expansion,* meaning that the ability of the steam to do work has been reduced. As the steam passes through the governor, a substantial amount of its pressure is converted to heat, because of friction in the governor valve.

The 300-psig steam next passes through the steam nozzle. This is an ordinary nozzle. It screws into a hole in the wall, which separates the steam chest from the *turbine case.* The nozzle is shaped to efficiently convert the pressure of the 300 psig to steam velocity. The pressure of the steam, as soon as it escapes from the steam nozzle, is already the same as the exhaust steam pressure (100 psig).

All the pressure energy lost by the steam in expanding from 300 to 100 psig is converted to velocity. This is called an *isoentropic* or *reversible expansion.* The term *isoentropic* is a thermodynamic expression meaning that the *entropy* of the steam has not increased. The term *reversible* in this context means that I could take the high-velocity, 100-psig, steam exhausting from the nozzle, run it backward through an identical nozzle, and convert the steam's kinetic energy back into 300-psig steam. An isoentropic expansion is also frictionless.

The high-velocity steam strikes the buckets around the rim of the turbine wheel shown in Fig. 17.1. Actually, these so-called buckets resemble blades or vanes. I imagine the term "bucket" is a carryover from the days of the water wheel. The velocity of the steam is now transferred to the spinning turbine wheel. If the turbine is running below its set speed, the speed controller causes the governor speed-control valve to open.

Calculating work available from motive steam

Dear engineering reader, please recover from your desk drawer, your *steam tables.* In the back, there will be a Mollier diagram for steam. Figure 17.2 is a representation of your Mollier diagram. We will use

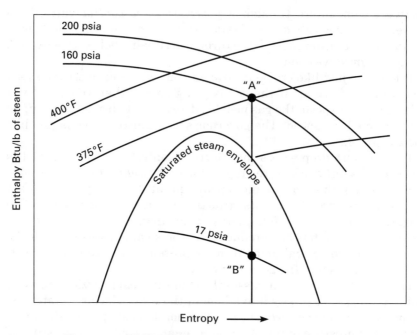

Figure 17.2   How to use a Mollier diagram.

this diagram to calculate the amount of work we can recover from steam with a turbine:

Step 1.   The motive steam conditions are 375°F and 160 psia, which determine point *A*. Note that the motive-steam pressure is the pressure in the steam chest; note also the supply pressure.

Step 2.   When steam passes through a turbine, it undergoes an isoentropic expansion. The work that the steam does in transferring its momentum to the turbine wheel exactly equals the shaft horsepower developed by the turbine. The entropy of the system is therefore constant. On this basis, extend a line through point *A* straight down the Mollier diagram. This line represents a constant entropy expansion.

Step 3.   The exhaust-steam pressure is 17 psia. The intersection of the isoentropic expansion line and 17-psia constant-pressure line together determine point *B* in Fig. 17.2.

Step 4.   Now measure the enthalpy difference in British thermal units per pound of steam between points *A* and *B* on the vertical (*y* axis) scale.

Step 5.   Divide your answer by 2500 Btu/hp. This is the fraction of a horsepower that each pound of steam can produce. However, as turbines are not 100 percent efficient, multiply the calculated horse-power by 0.90, to account for internal turbine inefficiencies called *windage losses.* Incidentally, the reciprocal of the calculated horse-power per pound of steam, is called the *water rate,* which is the pounds of motive steam needed to generate one horsepower worth of shaft work.

Step 6.   Multiply the horsepower per pound of steam value calculated in step 5 by the turbine steam flow, in pounds per hour. This is the total shaft work that appears at the turbine's coupling. This is the amount of horsepower that is available to spin a centrifugal pump.

Exhaust steam conditions

You may have noticed that point *B* in Fig. 17.2 is below the saturated-steam envelope line. Does this mean that the exhaust steam would have appreciable amounts of entrained water? Does this mean that the motive steam may sometimes partially condense inside a turbine? Answer—yes! Does this also mean that turbine exhaust steam may have too high a moisture content to be used in certain downstream ser-vices? Well, it is fine for reboilers, but it would be unsuitable for steam jets (see Chap. 16).

Looking at Fig. 17.2, see if you can agree with these statements:

- The higher the motive-steam superheat temperature, the drier the exhaust steam.

- The higher the exhaust-steam pressure, the drier the exhaust steam.

Lowering the exhaust-steam pressure always allows us to extract more work from each pound of steam. That is why we often exhaust steam to a condenser. But other than minimizing the exhaust-steam pressure, how else may we increase the amount of work that can be extracted from each pound of steam?

Horsepower valves

It is the velocity of the steam, impacting on the turbine wheel buckets, that causes the turbine to spin. If that is so, then the way to extract more work from each pound of steam is to increase the velocity of the steam as it escapes from the steam nozzle, shown in Fig. 17.1.

The escape, or exit, velocity of this steam is a function of the steam pressure in the steam chest. If we raise the pressure in the steam chest by 30 percent, then the velocity of the steam leaving the nozzle would go up by 30 percent as well. It is true that I could simply open up the governor, and reduce the pressure drop of the steam across the governor from 100 to 10 psi. This would raise the pressure in the steam chest from 300 to 390 psig. But I would also get 30 percent more pounds of steam flow through the nozzle. The turbine would spin a lot faster and develop a lot more horsepower. But this is not my objective.

What I wish to achieve is to maintain the same horsepower output from the turbine. But at the same time, I want to force open the governor speed-control valve, raise the pressure in the steam chest, but decrease the steam flow through the steam nozzle. The only way this can be done is to make the nozzle smaller.

We could shut down the turbine and unbolt the steam chest, to expose the *nozzle block,* which is the wall that separates the steam chest from the turbine case. We could unscrew the existing nozzle, and replace it with a smaller nozzle. A nozzle of 20 percent less diameter would reduce the nozzle cross-sectional area by 36 percent:

$$(1.00 - 0.20)^2 = 0.64 = 64\%$$

$$100\% - 64\% = 36\%$$

This procedure is called reducing the port size of the nozzle. It works fine, except that we have permanently derated the capacity of the turbine by 36 percent, and next month we might need this capacity.

Speed valves

Figure 17.3 shows a steam turbine with three, rather than one, nozzles. The single, largest, left-hand valve is called the *main nozzle.* It handles 60 percent of the motive-steam flow. Each of the two smaller nozzles handles 20 percent of the steam. These 20 percent nozzles can be plugged off by a device sometimes called either a *horsepower* valve, *jet* valve, *speed* valve, *star* (for the handle shape) valve, or *port* valve.

If we close off one of the two horsepower valves, steam flow into the turbine will drop—initially by 20%. The turbine will slow. This will cause the governor valve to open. The pressure drop across the governor will decrease. The pressure in the steam chest will rise. The flow of steam through the 60 percent port nozzle and the remaining 20 percent port nozzle will increase. The velocity of the steam striking the buckets will also increase. The turbine wheel will now come back up to its set point.

The net effect of this exercise will be to save not 20 percent of the motive steam, but 10 percent. The 20 percent reduction in nozzle area is partially offset by the opening of the governor valve. The inefficient, irreversible, isoenthalpic expansion and pressure drop across the governor speed control valve are reduced. The efficient, reversible, isoentropic expansion and pressure drop across the nozzles are increased.

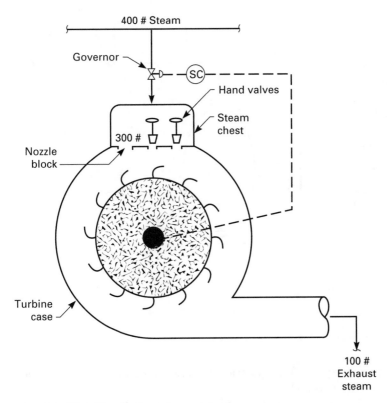

Figure 17.3 Use of hand valves can save steam.

If we now attempt to shut the second horsepower valve, we may be able to save a second 10 percent increment of steam. But, if the governor goes wide open, we will have to open this horsepower valve again. You see, once the governor is 100 percent open, we will have lost our ability to control the turbine's speed. This is no good, because we want the turbine to run at a constant speed.

Do not pinch on the horsepower valves. They must be left completely opened, or firmly closed. Leaving them in a partially open position will cut the valve seats and greatly reduce the efficiency of the turbine.

Selecting Optimum Turbine Speed

An ordinary American-type turbine is designed to run at 3600 rpm. Its overspeed trip will cut off the motive steam flow at about 3750 rpm. But the turbine can be run at any lower speed. There is usually a small knob on the left side of the governor-valve assembly, that is used to

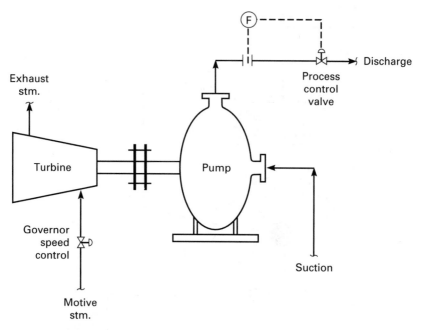

Figure 17.4   Selecting optimum turbine speed.

change the set speed. The question is, how do we know the best speed to operate the turbine?

Figure 17.4 shows a centrifugal pump, driven by a steam turbine. The correct operating speed for the pump and turbine is that speed that puts the process-control valve in a mostly open, but still controllable, position. As we slow the turbine to force open the process-control valve, the turbine's governor valve will close. Steam flow to the turbine will decline in accordance with *fan laws:*

$$\text{Shaft work} \sim (\text{speed of driver})^3$$

This means that if we drop the speed of the turbine by about 3 percent, motive-steam flow will decline by about 9 percent. However, there is an additional benefit.

Slowing a turbine closes the governor valve. This may now permit us to close an additional horsepower valve, without losing our flexibility to control the speed of the turbine. Closing that final horsepower valve will save us another 10 percent of steam.

Our overall objective is to wind up with both the process-control valve downstream of the pump and the governor speed-control valve in a mostly wide-open, but still controllable, position. To achieve this dual objective, we have to simultaneously:

- Optimize the turbine speed
- Optimize the number of horsepower valves that are closed

It is rather like solving two equations with two unknowns.

An old, but better, idea

The control scheme shown in Fig. 17.4 is certainly quite common. But is it the best? Figure 17.5 is a copy of the crude charge system in a now-defunct refinery in Port Arthur, Texas. I saw it in operation many years ago. It worked fine. The required flow of crude directly controls the governor. The turbine speed is then always at its optimum. The ΔP across the process-control valve is always zero, because there is no process-control valve. This design is a direct descendant of the original method of controlling the steam flow to pumps. The steam inlet valve was opened by the operator, so that the desired discharge flow was produced.

Why, then, do we need process-control valves, on the discharge of variable-speed-driven pumps? Why, indeed!

Steam rack

On larger steam turbines, we have automatically, rather than manually, operated horsepower valves. The mechanism that controls the movement of the horsepower valves, is called a *steam rack*. If you have

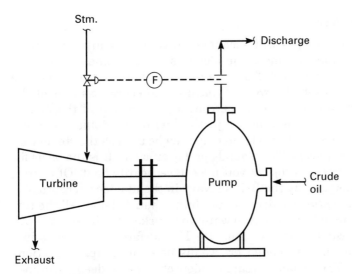

Figure 17.5 A better way to run a turbine.

a large steam turbine driving a compressor, you have likely seen such a steam rack.

A horizontal rod extending across the width of the turbine has a half-dozen cams fixed along its length. These cams lift and lower, in sequence, plungers attached to the horsepower valves. Watch these plungers in action. If you slow the compressor, here is what will happen:

- The governor valve will start to close.

- The steam chest pressure will fall.

- The steam rack will move, so as to close only one of its multiple horsepower valves.

- The steam chest pressure will rise, and the governor will open back up.

If you look on the local control panel of such a turbine, you will probably see four steam pressures displayed:

- Steam header pressure

- First-stage steam chest pressure

- Second-stage steam pressure

- Exhaust-steam pressure

These larger turbines often let down the steam in two or more stages. They are called *multistage turbines.* Their principle of operation is not significantly different from a simple single-stage machine.

Condensing turbines

Most of the turbines you will encounter in your work are called "topping," or *extraction,* turbines. The idea of such a turbine is to extract much of the potential work from the motive steam, and then use the exhaust steam to reboil towers. Typically, the energy content of the exhaust steam is only 10 to 20 percent less than that of the motive steam. That is the calculation we just did with the Mollier diagram. The rest of the energy of the steam may then be used as the steam condenses, to reboil towers. This sounds pretty efficient. It is the basis for the new *cogeneration* projects you may have heard about. Of course, this system was used by the British Navy in the nineteenth century.

The *condensing turbine* does not produce exhaust steam. All the turbine exhaust steam is turned into water in a surface condenser. We will study surface condensers in Chap. 18. The surface condenser is just like the sort of vacuum condensers we discussed in Chap. 16, sections on steam jets. The exhaust-steam condenses under a deep vacuum—typically 76 mm Hg, or 0.1 atm. Basically, then, a condensing steam turbine loses all the potential benefit of using its exhaust steam to reboil towers. But for what purpose?

Let's say a turbine was using 400-psig (i.e., 415-psia) steam exhausting to a 30-psig (i.e., 45-psia) steam header. The work generated was 2000 hp. If we had condensed the steam at 76 mm Hg (i.e., 1.5 psia), the work generated would be 4000 hp. In the first case the ΔP of the steam was 370 psi (415 − 45 = 370). In the second case, the ΔP of the steam is 413.5 psi (415 − 1.5 = 413.5).

If the ΔP increases by only 12 percent, why does the amount of work that is extracted from each pound of steam double? The answer is that it is not the pressure of the steam that drives the turbine. It is the velocity of the steam impacting on the turbine wheel buckets.

The velocity of the steam escaping from the steam nozzle (see Fig. 17.1) is 25 ft/s, when the exhaust pressure is 30 psig (45 psia). What, then, would be the velocity of the steam if the exhaust pressure were 76 mm Hg (1.5 psia)?

$$\frac{25 \text{ ft/s} \times 45 \text{ psia}}{1.5 \text{ psia}} = 750 \text{ ft/s}$$

No wonder the work we can extract from a condensing steam turbine is so much greater than with an extraction, or topping, turbine. The most efficient condensing turbines may convert 30 percent of the steam's energy into work. These turbines use 1500-psig steam, and exhaust to a surface condenser at 50 mm Hg. The most efficient machine ever built is the high-bypass, radial-flow jet turbine engine, commonly used in commercial aircraft.

I rather like to adjust steam turbines. It reminds me of Professor Peterson, my thermodynamics instructor. It makes me think that the 4 years I spent at Cooper Union was time well spent. It is just nice to see how the Second Law of Thermodynamics functions in the real world.

18

Surface Condensers

The Condensing
Steam Turbine

More of the world's energy is consumed in surface condensers than for any other single use. Even at the very start of the industrial revolution, the father of the surface condenser, the *barometric condenser,* consumed huge amounts of heat.

The central idea of our industrialized society is to have machines do the work formerly done by humans or animals. The technical problem that kicked off the industrial revolution was flooding in the tin mines of Cornwall, a region in southern Britain. To work the mines and extract the valuable tin ore, steam-driven reciprocating water pumps had been constructed. These plunger-type pumps were moved up and down by the famous beam engines, and these reciprocating engines were powered by steam, thanks to the efforts of Thomas Newcomen, in 1712.

The motive steam for these reciprocating engines was charged into a steam cylinder. The piston inside the cylinder was pushed up by the expanding steam. The piston then lifted the beam attached to the reciprocating pump. Cold water was admitted next and jetted into the steam cylinder. The cold water absorbed the latent-heat content of the steam. The steam pressure inside the cylinder dropped, and the piston dropped. This pulled the beam down. It was the up-and-down movement of the beam that powered the reciprocating pump. The structures, beams, foundations, and bits and pieces of ancient machinery are scattered all across Cornwall. (Caution: The tin mine entrances are rather obscure, overgrown with brambles, and your author more or less fell into one. If you choose to explore the mining areas of Cornwall, be very careful as many of these shafts are hundreds of feet deep. There are numerous stories about people and livestock slipping into these disused shafts and being drowned in the accumulated water or simply falling to their deaths down the mine shafts.)

There was a big problem with the initial beam engine design. When the cold water was admitted into the steam cylinder, the water not only absorbed the latent heat of the condensing steam but also cooled the iron walls of the steam cylinder. Then, when the next charge of steam was admitted to the cylinder from the boiler, a lot of the steam's heat was wasted in reheating the iron walls of the cylinder. Then, again, an awful lot of cold water was wasted in cooling the steam cylinder each time the motive steam had to be condensed.

All this wasted a tremendous amount of coal. Actually, only 1 to 2 percent of the energy of the coal was converted into useful work. Quite suddenly, the steam engine was revolutionized. Its efficiency was increased by a factor of 10. This was all due to the innovations of James Watt, who invented the external barometric condenser, in the late 1760s.

The Second Law of Thermodynamics

Steam power was fully developed before the introduction of the science of thermodynamics. The steam engine was designed and built by ordinary working people, such as Mr. Newcomen and Mr. Watt. Mr. Watt's invention is illustrated in Fig. 18.1.

Rather than cooling and condensing the steam in the cylinder, the steam was exhausted to an external condenser. In this external condenser, the exhaust steam was efficiently contacted with the cold water. This external or barometric condenser rather looks and performs like the deaerator we discussed in Chap. 15. The external condenser obviously achieved Mr. Watt's original objective. He could condense the steam without cooling the cylinder. But the barometric external condenser was found to have an even more important attribute. Let me explain.

The boilers in those days were not pressure vessels. They were constructed from sheets of wrought iron and assembled by riveting and hammering the seams. At best, they could hold a few atmospheres of steam pressure. The low-pressure steam generated did not really push the beam and piston up. The beam was pulled up by a heavy weight attached to the far end of the beam. This also pulled the piston up.

Well, if the steam did not really push the piston up, what did the steam do? Answer—the condensing steam pulled the piston back down. As the steam condensed, it created an area of very low pressure below the piston (see Fig. 18.1). This low pressure sucked the piston down.

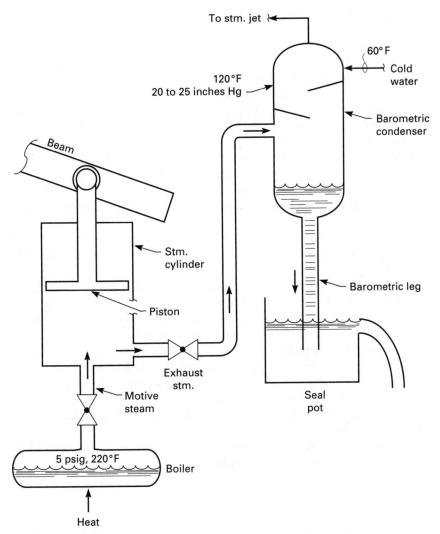

Figure 18.1 Barometric condenser; improved efficiency of the legendary beam engine.

The colder the temperature at which the steam condensed, the lower the pressure at which the steam condensed.

The lower the pressure in the steam cylinder, the more forcefully the piston was drawn down. And the more forcefully the piston was sucked down, the more work the beam engine could do with the same amount of coal consumed.

It is easy to see how the barometric condenser could condense the exhaust steam more efficiently than periodically squirting water into the steam cylinder. The barometric condensers could absorb the latent heat of condensation of the steam at temperatures of 120°F or less. Water condenses at a pressure of 25 inches Hg at this temperature. This extremely low vacuum sucked the piston down more forcefully because of the greater differential pressure across the cylinder. In other words, on top of the cylinder, there was atmospheric pressure; below the cylinder, there was the pressure in the barometric condenser.

Two problems arose with the use of the barometric condenser. First, if the condenser operated at subatmospheric pressure, how can the water be drained out of the condenser? That is easy. Set the condenser on a hill 34 ft high. Then drain the water down, through a barometric leg, to a seal pot. The pressure which a column of water 34 ft high exerts is equal to one atmosphere, or one bar. Hence the term, barometric "leg."

The second problem was air leaks. Air drawn into the system, would build up in the condenser. This noncondensable vapor was drawn off by using a steam jet.

Certainly, if we could generate steam at a higher pressure and temperature in the boiler, we could push the piston in Fig. 18.1 up with greater force. Thus thought Richard Trevithick of Cornwall, who pioneered the use of high-pressure steam in the 1790s. And as the mechanical design of boilers improved the last 200 years, this was done, and we can now push up the beam with greater force. As a result, the amount of work that was extracted from steam more than doubled. But it was James Watt, working alone, repairing a model of a Newcomen engine,[1] who made the big leap forward in improving the efficiency of the steam engine.

And what, dear reader, does all this have to do with the Second Law of Thermodynamics? This law states

■ $W = (\Delta H)(T_2 - T_1)$ (18.1)

where

■ W = amount of work that can be extracted from the motive steam

■ ΔH = enthalpy of the motive steam, minus the enthalpy of the condensed water

■ T_2 = temperature at which the steam is generated in the boiler

■ T_1 = temperature at which the steam is condensed in the barometric condenser

According, then, to the Second Law of Thermodynamics, Mr. Watt lowered T_1. But, of course, the professors who worked out these laws were just formalizing the discoveries that practical working people had made a 100 years before their time, using common sense and craftsmanship.

The surface condenser

There is another problem with the barometric condenser that did not become apparent at first. When the British Navy decided to convert from sail to steam, this problem was immediately obvious. While steam can be generated from seawater, it is far better to use freshwater, especially if one wishes to generate high-temperature, high-pressure steam. And as freshwater supplies are limited at sea, it would be great if the condensed steam could be recycled to the boilers. But the cooling-water supply to the barometric condensers was naturally seawater which mixed with the steam condensate.

The solution is straightforward. Do not condense the steam by direct contact with cold water, as is done in the barometric condenser. Condense the steam by indirect contact, with the cold surface of the tubes in a shell-and-tube condenser. Hence the name *surface condenser,* a sketch of which is shown in Fig. 18.2. Compare Fig. 18.1 with the surface condenser. Is there really much difference? Other than recovering clean steam condensate for reuse, there is no difference at all. I last used a surface condenser in 1976, on a sulfuric acid plant reactor feed gas booster blower, and it worked just fine.

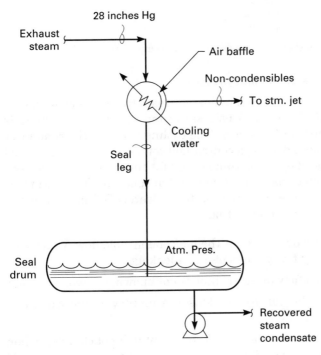

Figure 18.2 Surface condenser.

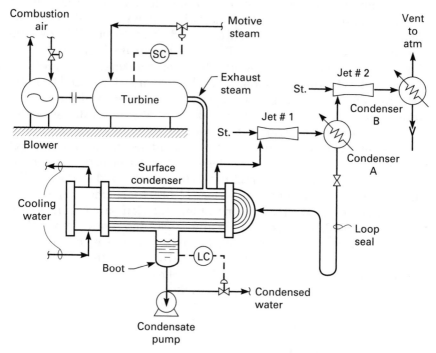

Figure 18.3 Condensing steam trubine driving an air compressor.

Using the Second Law of Thermodynamics

The motive-steam supply to a condensing steam turbine such as shown in Fig. 18.3 is 360°F and 150 psig saturated steam. The turbine is exhausting to a surface condenser. The cooling water to the condenser is 92°F. The turbine is driving a centrifugal compressor. The calculated horsepower produced by the turbine is 10,000 bhp (brake horsepower). Bill Duvall, your supervisor, has told you that colder 62°F well water is to be substituted for the 92°F cooling-tower water. Bill has given you the following additional information:

■ The steam jets are oversized for the noncondensable flow, which consists of only a very few pounds of air in-leakage.

■ The motive-steam flow to the turbine is not known, but won't change.

■ The pressure in the surface condenser is unknown, and cannot be measured.

■ The cooling-water flow rate is not known, but will not change, either. The water will just be colder.

- The efficiency of the turbine and compressor is not known, but is presumed to remain constant.

Bill has asked you to calculate the new compression horsepower output from the compressor. Using Eq. (18.1), we note

- Compression work W is proportional to horsepower.
- ΔH will go up from its prior value a little because the enthalpy of the condensed steam will be lower (because it's colder).
- As the cooling-water supply is 30°F lower, we will assume that the condensation temperature T_1 in Eq. (18.1) is reduced by 30°F.
- The enthalpy difference between 150-psig saturated steam, at 360°F, and steam condensed under a good vacuum, is roughly 1000 Btu/lb.
- The enthalpy reduction of condensing the steam at a 30°F lower temperature will increase ΔH by 30 Btu/lb.
- ΔH in Eq. (18.1) thus increases by 3 percent.
- T_2, the temperature of the motive steam, is always 360°F. With 92°F cooling water, we will assume that T_1 is 120°F.
- $(T_2 - T_1)$, with 92°F cooling water, is (360°F−120°F) = 240°F.
- T_1, with 62°F cooling water, is 30°F cooler than T_1, with 92°F cooling water; that is, the new T_1 is (120°F − 30°F) = 90°F.
- $(T_2 - T_1)$, with 62°F cooling water, is then (360°F − 90°F) = 270°F.
- With the cooler well water substituted for the warmer cooling-tower water $(T_2 - T_1)$ has increased by

$$\frac{(270°F - 240°F)}{(240°F)} = 12.5\%$$

Combining the ΔH effect of 3 percent with the larger $(T_2 - T_1)$ effect of 12.5 percent in Eq. (18.1) results in a compression horsepower increase of 16 percent or about 11,600 bhp total. If you wish, work through the same problem but use 5-psig saturated steam, which was used to power steam engines in the eighteenth century, rather than the 150-psig saturated steam we used in our current example. The answer will illustrate why Mr. Watt is still well remembered in Cornwall along with justly revered Mr. Trevithick.

Surface Condenser Problems

In Chap. 16 we reviewed several problems pertaining to steam jet precondenser and intercondenser problems. The surface condensers, which serve condensing steam turbines, are subject to all the same problems, and a lot more. A standard surface condenser package with an associated two-stage jet system is shown in Fig. 18.3. By way of summarizing many of the problems which occur with this sort of equipment, I will relate my trials and tribulations with K-805, an auxiliary combustion air blower at the Good Hope Refinery.

The problem with this new air blower, was that we could not bring the turbine speed above its *critical speed*. The critical speed of a turbine is stamped on the manufacturer's nameplate. Turbines are typically run well above their critical speed. If, for some reason, a turbine is run close to its critical speed, it will experience uncontrolled vibrations and self-destruct.

For a surface condenser to work properly, noncondensable vapors must be sucked out of the shell side. This is done with a two-stage jet system, as shown in Fig. 18.3. When first commissioned, the jets were unable to pull a good vacuum. Moreover, water periodically blew out of the atmospheric vent. I found, after considerable investigation, that the condensate drain line from the final condenser was plugged.

I directed the maintenance crew to disassemble and clean the drain lines from both the final condenser (B), and the primary jet discharge condenser (A). Unfortunately, they failed to reassemble the *loop seal* from condenser A. But what is the purpose of this loop seal?

The pressure in condenser A is greater than that in the surface condenser, and less than that in the final condenser (condenser B). This means that condenser A is operating at vacuum conditions. This prevents the condensed steam formed in condenser A from draining out to atmospheric pressure, unless the condenser is elevated by 10 to 15 ft. To avoid this problem, the condensate is drained back to the lower-pressure surface condenser. To prevent blowing the noncondensable vapors back to the surface condenser as well, a loop seal is required. The height of this loop seal must be greater than the difference in pressure (expressed in ft of water) between the surface condenser and the primary jet discharge condenser (condenser A).

With the loop seal gone, the noncondensable vapors simply circulate around and around, through the primary jet, but no substantial vacuum in the surface condenser can be developed.

Having replaced the loop seal piping, (some units use a steam trap instead of this loop seal), I started steam flow to the turbine. But the vacuum in the surface condenser, which had started out at an excellent 27 in Hg, slipped down to 14 in Hg. This loss in vacuum increased the backpressure in the turbine case. The higher pressure in the turbine case reduced the velocity of the steam striking the buckets on the turbine wheel, which reduced the amount of work that could be extracted from each pound of steam.

For practice, pull out your Mollier diagram. If the motive steam is 400 psig and 650°F, what is the effect of reducing the vacuum in the surface condenser from 27 to 14 in Hg? Answer: 13 percent loss in horsepower (see Chap. 17).

The turbine began to slow. It slowed to its critical speed, and began to vibrate. Before shutting down the turbine, to avoid damage due to the vibrations, I noted the following:

- The temperature of the turbine vapor exhaust, to the primary jet had increased from 125 to 175°F.

- The temperature of the condensate draining from the surface condenser had decreased from 125 to 100°F.

- The condensate pump was cavitating, as indicated by an erratically low discharge pressure.

The increase in the vapor outlet temperature from a condenser, as compared to a decrease in the temperature of the condensate from the same condenser, is a sure sign of condensate backup. The condensate is covering some of the tubes in the surface condenser. This subcools the condensate and does no harm.

However, the number of tubes exposed to the condensing steam is also reduced. This forces the steam to condense at a higher temperature (as discussed in Chaps. 8 and 13). In effect, the condensate backup has reduced the surface area of the condenser, available to condense the steam. The higher the condensation temperature of the steam, the higher the condensation pressure of the steam. Just like the deaerator I described in Chap. 15.

Take a look at Fig. 18.3. It is the vapor outlet temperature of the surface condenser, rather than the condensate outlet temperature of the surface condenser, that determines the real condensing temperature and pressure of the exhaust steam.

Condensate pumps serving surface condensers have a common problem. Their suction is under a vacuum. For example, let's assume the following for Fig. 18.3:

- The surface condenser pressure = 27 in Hg.

- The condensate water level in the boot is 11 ft above the suction of the pump.

- One inch of mercury (1 in Hg) is equal to a head of water of 1.1 ft (this is a good rule of thumb worth remembering).

The pressure at the suction of the condensate pump is then

$$27 \text{ in Hg} - 11 \text{ ft H}_2\text{O} \times \frac{1 \text{ in Hg}}{1.1 \text{ ft H}_2\text{O}} = 17 \text{ in Hg}$$

Often, centrifugal pumps develop seal leaks. If the suction of a centrifugal pump is under a vacuum, air will be drawn into the pump through the leaking seal. The pump's capacity will be severely reduced. To stop the suspected air leak, I sprayed water from a hose over the pump's seal. Now, instead of sucking air, the leaking seal drew in cold water. As a result

- The cavitation of the condensate pump stopped.

- The high water level in the boot was pulled down.

- The condensate outlet temperature increased.

- The vapor (or noncondensable) outlet temperature decreased.

- The vacuum in the surface condenser was restored.

- And the turbine speed came back up, well above its critical speed.

But not for long. After 15 min of operation, the turbine speed slipped back down. Once again, I had lost a lot of vacuum in the surface condenser. Once again, the vapor outlet temperature had dramatically increased. But this time, the condensate outlet temperature had also increased. What was my new problem?

I now observed that the surface condenser cooling-water outlet temperature had increased from 100 to 135°F. This is a sign of loss of cooling-water flow. As none of the other water coolers in the plant had been affected, I concluded that the cooling-water inlet to my surface condenser was partly plugged.

I had the front endplate on the cooling water side of the surface condenser (called the *channel head cover*) removed. Most of the tube inlets in the *channel head tubesheet* were plugged with crayfish (but in Louisiana where this story is set everyone calls these little creatures, crawfish).

The offending wildlife were removed. The condenser was reassembled. The motive steam was started to the turbine. Both the turbine and the air blower were running well above critical speed. We lined the flow of combustion air up to the combustion chamber. Everything was finally going my way, except for one minor problem.

The governor speed-control valve, shown in Fig. 18.3, was 100 percent open. The plant's boiler house was not sending us the proper pressure steam. I was supposed to be getting 460-psig steam, but was only receiving 400-psig steam. This reduced the steam flow through the nozzles in the turbine's steam chest by 15 percent. The operators at the power station assured me that the problem was temporary. The normal 460-psig steam pressure would be restored by morning.

Ladies and gentlemen, it is not a good idea to run a turbine with the governor speed-control valve wide open. Why? Because you no longer have any speed control. And the turbine speed is then free to wander. The rest of this story is pure philosophy.

During the evening, my operators decided to increase the combustion airflow from the blower or air compressor. This is done by opening the suction valve to the blower. Naturally, it requires more work to compress more air. But the turbine could not produce any more work or horsepower because the steam turbine's governor speed-control valve was already 100 percent open!

So, the turbine slowed down. And what was the only possible speed that it could slow down to? Why, the critical speed, of course. The turbine and blower began to vibrate. The bearings were damaged. The turbine's rotor became unbalanced. A rotating element on the air blower touched a stationary component in the blower's case. The stationary component broke off and wrecked the blower. That was the end of K-805.

By what law of nature was the turbine forced to slow exactly to its critical speed? You see, dear reader, life is perverse. And if anything bad can happen, it is going to happen to me.

Noncondensable load

The gas that accumulates inside the surface condenser is called the *noncondensable load to the steam jets.* Some of the noncondensable load consists of CO_2 accidentally produced when the boiler feedwater is vaporized into steam. Air leaks through piping flanges and valves are other sources of noncondensable vapors. But the largest source of non-condensable vapors is often air drawn into the turbine case, through the shaft's mechanical seals. To minimize this source of leaks, 2 or 3 psig of steam pressure is ordinarily maintained around the seals. However, as the turbine's shaft seals deteriorate, air in-leakage problems can overwhelm the jet capacity. This will cause a loss of vacuum in the surface condenser.

If vacuum in the surface condenser is bad, there are two possible causes. Either the jets are at fault, or the surface condenser is at fault.

1. *Jet problems.* These include low motive-steam pressure, excess wear on the steam nozzles, high condenser backpressure, and air leaks that exceed the jet's capacity. To determine whether a poor vacuum in a surface condenser is due to such jet problems, consult the chart shown in Fig. 18.4. Measure the surface condenser vapor outlet temperature and pressure. Plot the point on the chart. If this point is some-

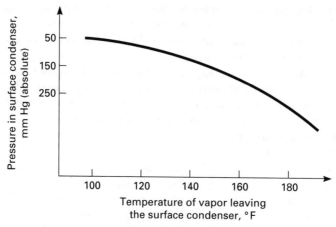

Figure 18.4 Vapor pressure of water under vacuum.

what below the curve, your surface condenser's loss of vacuum is due, at least in part, to jet deficiencies.

2. *Surface condenser problems.* These include undersized surface condenser area, water-side fouling, lack of water flow, condensate backup, and excessive cooling-water inlet temperature. To determine whether a poor vacuum in a surface condenser is due to such heat-transfer problems, plot the same point on the chart shown in Fig. 18.4. If this point is on or slightly below the curve, it is poor heat transfer in the surface condenser itself that is hurting the vacuum.

The curve in Fig. 18.4 also represents the best possible vacuum that can be obtained in any surface condenser. The majority of surface condensers I have seen do operate right on the curve. Condensers operating below the curve are typically suffering from air in-leakage through the turbine shaft seals.

Function of the final condenser

We discussed before that the drain from the final condenser shown in Fig. 18.3 had plugged. Rather than unplugging the drain, could we have simply disconnected the final condenser (condenser B), and vented the discharge from the secondary jet (jet 2) to the atmosphere? Would this have helped or hurt the vacuum in the surface condenser?

The final condenser hurts the upstream vacuum. The final condenser increases the discharge pressure from the secondary jet and thus makes the jet system work slightly harder to expel the noncondensable gas load.

What, then, is the true function of the final condenser? Well, if the tiny amount of condensed steam is not needed, the final condenser serves no function at all. It may safely be discarded. Why, then, do surface condensers come with final condensers? It is just a convention that, for most plants, makes no particular sense. It is really just a hangover from the design to conserve freshwater on the old navy ships.

Surface Condenser Heat-Transfer Coefficients

Heat-transfer coefficients in this book have the units of Btu/[(h)(ft^2)(°F)], where the ft^2 term refers to the surface area of the surface condenser. The °F term refers to the condensing steam temperature, minus the average tube-side cooling-water temperature.

Most unfortunately, an optimistic correlation for heat-transfer coefficients for surface condensers has become widely disseminated in several books devoted to heat transfer. This correlation predicts heat-transfer coefficients, for clean condensers, of about 650, when the water-side velocity is about 6 ft/s. Use of this correlation has led to some extremely serious problems, with which your author is intimately acquainted.

The correct heat-transfer coefficient for a surface condenser, with a water-side velocity of 6 ft/s, is about 200 to 240. Including an allowance for fouling, we suggest you use 140 to 160 overall heat-transfer coefficient, for steam surface condensers. While I have observed clean coefficients approaching 400, I would not count on maintaining 400 for several years on an industrial surface condenser.

Reference

1. B. Trinder, ed., *The Blackwell Encyclopedia of Industrial Archaeology,* Blackwell, London, 1992.

19

Shell-and-Tube Heat Exchangers

Heat-Transfer Fouling Resistance

After distillation, heat transfer is the most important operation in a process plant. Most of the heat transfer in chemical plants and petroleum refineries takes place in shell-and-tube heat exchangers. The surface condenser we discussed in Chap. 18 is an example of a shell-and-tube heat exchanger.

A wide variety of heat exchangers are available, some of which you may have seen: plate, spiral, and coil—to name just three. But 99 percent of the heat exchangers I have worked with are ordinary shell-and-tube exchangers, the design of which has not changed since the 1920s.

Before considering the process aspects of heat transfer, let's look at the mechanical components of the heat exchanger shown in Fig. 19.1. This shell-and-tube heat exchanger is actually a compromise between four aspects of heat-exchanger design:

- Allowance for thermal expansion
- Efficient heat transfer
- Ease of cleaning
- Mechanical robustness

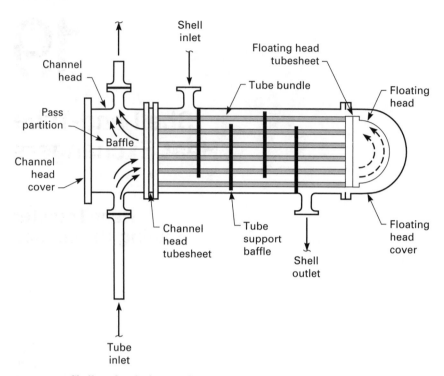

Figure 19.1 Shell- and -tube heat exchanger.

Allowing for Thermal Expansion

Referring to Fig. 19.1, we can see how a *floating-head exchanger* works. The tube-side flow enters the bottom of the *channel head*. This assumes the cold fluid to be on the tube side. The cold fluid may be on the shell side or the tube side of an exchanger. The convention is to put the cold fluid nozzle on the bottom of the exchanger. Sometimes this is necessary. Sometimes it does not matter, but it is still the convention.

Inside the exchanger's channel head, we have the *pass partition baffle,* which divides the channel head into two equal portions. This baffle forces the total flow only through the bottom half of the tubes. The tubes themselves are pipes of either $\frac{3}{4}$ or 1 in OD (outside diameter). The front end of each tube is slipped into a slightly larger hole drilled into the *channel head tubesheet.* This tubesheet is a disk about 2 in thick, slightly larger than the inner diameter of the shell (shell ID).

The tubes are firmly attached to the tube sheet by "rolling." After a tube is pushed into the tubesheet, a tapered tool is inserted into the open end of the tube, and forcefully rotated. The tube's diameter is thus slightly expanded. While rolling is quite effective in sealing the tube inside the tubesheet, rolls have been known to leak.

The tube-side fluid now flows into the *floating head,* which acts as a return header for the tubes. The tube-side flow makes a 180° turn and flows back through the top half of the *floating-head tubesheet.* The floating head is firmly attached to the floating-head tubesheet. But why is it that one end of the tubes must be left free to float? The reason is thermal expansion—or, more precisely, the differential rate of thermal expansion between the tubes and the shell.

Not all shell and tube exchangers have a floating head. Many exchangers have individual U bends for each tube. Then, each of the U bends functions like a mini–floating head for each tube.

The requirement to leave one end of the tubes free to float creates a rather unpleasant process problem. The most efficient way to transfer heat between two fluids is to have true countercurrent flow. For a shell-and-tube exchanger, this means that the shell-side fluid and the tube-side fluid must flow through the exchanger in opposite directions. When calculating the *log-mean temperature driving force* (LMTD), an engineer assumes true countercurrent flow between the hot fluid and the cold fluid.

But this is not the case with a floating-head exchanger. The tube-side fluid reverses direction in the floating head. It has to. There is no way to attach the tube-side outlet nozzle to the floating head. It is a mechanical impossibility. So we bring the tube-side fluid back to the top half of the channel head. So, half of the tubes are in countercurrent flow with the shell-side flow. And that is good. But the other half of the tubes are in concurrent flow with the shell-side flow. And that is bad.

When calculating the LMTD for such a floating-head exchanger, the engineer has to apply the *F*-factor derating coefficient to reflect the loss in heat-transfer efficiency because of the floating head. A typical exchanger might lose 5 to 30 percent of its capacity because of nontrue countercurrent flow.

Designing to allow for thermal expansion

One-pass tube-side exchangers. I have not been completely accurate in the preceding discussion. Many exchangers are designed to bring the tube-side fluid in one end of the shell and out the other side. This is a true countercurrent arrangement. To provide for thermal expansion in such exchangers, an *expansion joint* is provided. These expansion joints are prone to leaking. They represent inferior mechanical engineering practice.

For moderate-temperature and low-pressure condensers, such as vacuum surface condensers (see Chap. 18), a single pass on the tube side is not uncommon. These exchangers are typically fixed-tubesheet designs. Such exchangers are fine from a purely process point of view. However, there is no practical way to disassemble the exchanger to clean outside (i.e., the shell side) of the tubes. This inability to clean the shell side frequently leads to tremendous loss of efficiency after the condensers foul.

Shell-side flow.   The hot shell-side flow enters the exchanger, as shown in Fig. 19.1, through the top inlet nozzle. Not shown on this sketch is the *impingement plate,* which is simply a square piece of metal, somewhat larger than the inlet nozzle. Its function is to protect the tubes from the erosive velocity of the shell-side feed. The plate lies across the upper row of tubes.

The four *tube support baffles* shown in this exchanger serve a dual function:

- They serve to support the tubes.
- More importantly, they promote high *cross-flow velocity.*

The concept of cross-flow velocity is quite important in understanding how heat exchangers work. This concept is related to a flow phenomenon called *vortex shedding.* Perhaps you have seen a wire quivering in the wind. What causes the wire to vibrate with such energy?

When a fluid such as air or water flows perpendicularly across a wire or tube, vortices, such as those shown in Fig. 19.2, are created. The resulting turbulence that forms behind a wire will cause the wire to

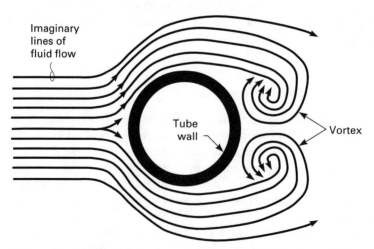

Figure 19.2 Liquid flow across a tube creates vortices and turbulence, thus improving heat transfer.

vibrate. The turbulence that forms behind a tube in a heat exchanger promotes good heat transfer.

Turbulence always encourages good sensible-heat transfer. The greater the velocity, the more violent the turbulence. For example, I recall a −40°F day in Fort McMurray, Alberta. The air is quite still, and I am comfortable. Suddenly the wind comes up. The rate of sensible-heat transfer from my body to the ambient air increases tenfold. Why does this happen?

Our bodies are always surrounded by a thin layer or film of stagnant air. This film is like a layer of insulation. It retards heat transfer between our skin and any surrounding fluid. Movement of the fluid causes turbulence. The turbulence disturbs the film and reduces the film's resistance to heat transfer.

To encourage vortex shedding and turbulence on the shell side of a heat exchanger, we must increase the cross-flow velocity. To calculate the cross-flow velocity, we proceed as follows:

1. Find the distance between the adjacent tube support baffles, in inches.

2. Count the number of tubes at the edge of the tube support baffle. As shown in Fig. 19.3, this would be seven.

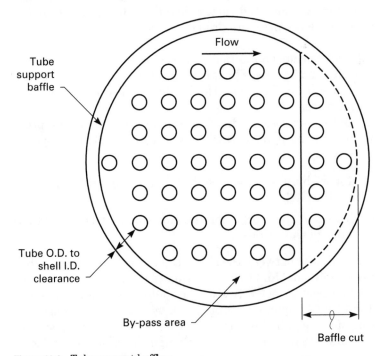

Figure 19.3   Tube support baffle.

3. Multiply the number of tubes by the space between each tube, in inches.

4. Multiply the inches measured in step 1 by the inches calculated in step 3.

5. Divide by 144 to obtain the shell-side cross-flow area in square feet.

6. Divide the pounds per second of shell-side flow by the fluid density in pounds per cubic foot. This will give you the volumetric flow in ft^3/s.

7. Divide the volume calculated in step 6 by the area calculated in step 5. This is the cross-flow velocity.

A good cross-flow velocity for water is 3 to 5 ft/s. For fluids other than water, a reasonable cross-flow velocity, in feet per second, is

$$\frac{30}{(\text{Density})^{1/2}}$$

where the *density* term is the density of the shell-side fluid, in pounds per cubic foot (lb/ft^3).

The baffle cut, shown in Fig. 19.3, is usually about 20 to 30 percent of the diameter of the baffle. The smaller the baffle cut, the more perpendicular the flow across the tubes. Perpendicular flow encourages desirable cross-flow velocity and vortex shedding. But a smaller baffle cut will also increase the pressure drop on the shell side.

Effect of shell-side pressure drop.   Reducing the baffle spacing increases cross-flow velocity and improves heat transfer. But it also increases the shell-side pressure drop. Reducing the baffle cut also improves heat transfer, but increases ΔP.

Too much shell-side pressure drop can create a problem. The problem is flow through the bypass area shown in Fig. 19.3. This bypass area is caused by two factors:

■ The tube support baffles must have a diameter somewhat smaller than the ID of the shell.

■ The holes drilled in the baffles for the tubes cannot be drilled too close to the edge of the baffles.

The gap thus created between the shell ID and the outer row of tubes will permit the shell-side fluid to bypass around the tubes. This is obviously very bad for heat transfer. And as the shell-side ΔP increases, the percent of fluid that is squeezed through the bypass area increases.

If the baffle spacing gets too small, the shell-side heat-transfer rate will actually worsen. This happens even though the cross-flow velocity increases. What can be done to correct this problem?

Shell-side seal strips. See if you cannot find an old tube bundle lying around your plant. Many such bundles have pairs of metal strips set around the edge of the tube bundle. These metal strips are typically ¼ in thick and 4 in wide. They extend down the length of the tubes. As seen in Fig. 19.4, they are inserted in grooves cut in the tube support baffles. These seal strips often increase heat-transfer efficiency by 5 to 10 percent.

The function of seal strips is to interfere with, and hence reduce, the fluid flow through the bypass area. Often, one pair of seal strips is used for every 18 in of shell ID (inner diameter). These seal strips encourage good shell-side cross-flow velocity and also help reduce localized fouling, caused by low velocity.

Please note that the tube support baffles are normally installed with the baffle cut in a vertical position. I have shown the baffle cut in Fig. 19.1 in a horizontal position for clarity. Baffles are installed vertically in most exchangers to reduce buildup of sludges in the bottom of the shell and avoid trapping vapors in the top of the shell.

Terminal-tube velocity. Have you ever seen a heat-exchanger tube bundle pulled? Perhaps many of the tubes were bent and twisted like partly cooked spaghetti. This distortion could not be very good for the

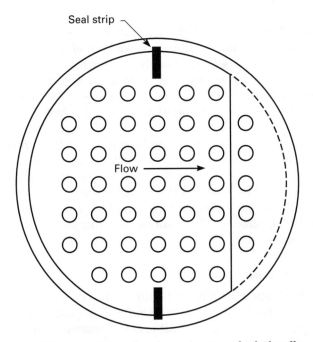

Figure 19.4 Seal strips reduce bypassing around tube bundle.

shell-side heat transfer. The cause of the bent tubes is called *terminal-tube velocity*. Let me explain.

Let's assume we have a new heat exchanger. The fluid on the tube side is crude oil. Crude oil, like many process fluids, will eventually foul and plug heat-exchanger tubes. The tendency to lay down fouling deposits is accelerated by

- Low velocity
- High temperature
- Dirt, salts, corrosion products, and other particulates

The crude oil is being heated by a hot-oil stream circulating through the exchanger's shell. Even though the exchanger is new, the flow of crude through all the tubes is not equal. Some of the tubes naturally run somewhat slower, and in some of the tubes, crude oil will run faster.

Those tubes that start running slower tend to get hotter. As the crude oil in these tubes gets hotter, the growth of fouling deposits accelerates. As the buildup of deposits in the slower tubes increases, the flow through these tubes is further restricted. The lower velocity increases fouling rates and restricts flow. This also increases the temperature of the slower tubes, and accelerates the rate of fouling. These tubes continue to plug and foul, and run slower, until crude-oil flow is essentially lost.

Meanwhile, the flow of crude oil through those tubes that started out running faster increases. The crude-oil flow is backed out of the slower tubes by the fouling deposits. The extra flow increases the velocity in the tubes that initially ran faster. The higher velocities retard the buildup of fouling deposits. Also, the greater flow keeps these tubes cooler. This also discourages the accumulation of deposits inside the tubes.

To summarize, some of the tubes in the bundle continue to foul until they plug off. These tubes get hotter until they reach the temperature of the hot oil circulating through the shell. Other tubes continue to receive more and more flow. The velocity in these tubes increases to the point where the rate of fouling becomes inconsequential. This velocity is called the *terminal-tube velocity*. These tubes may run 50 to 150°F cooler than the plugged tubes.

Differential rates of thermal expansion

It is quite true that the floating head permits differential rates of thermal expansion between the shell-and-tube bundle of an exchanger. However, the floating head cannot permit differential rates of thermal expansion between individual tubes.

The hotter, fouled tubes must grow. But their horizontal expansion is constrained by the cleaner, colder tubes. since the colder tubes do not allow the hotter tubes to grow, the hot tubes bend. This, then, is the origin of the twisted tubes we see when an improperly designed tube bundle is pulled from its shell during a turnaround.

The tendency to foul tubes could best be eliminated by eliminating salts, wax, particulates, corrosion products, polymers, free radicals, and all the myriad of other factors that contribute to fouling. Possibly— but not likely.

The temperature extremes of a process may occasionally be moderated. For example, we desuperheat steam to an amine regenerator reboiler, to slow degradation and fouling, with the heat-sensitive amine. We avoid heating di(ethylene glycol) above 360°F, to minimize fouling in our glycol dehydration unit water strippers. But in general, our latitude to manipulate process temperatures is very limited.

On the other hand, we can, during the design of an exchanger, select a high velocity to combat fouling. For example, let's consider Fig. 19.5.

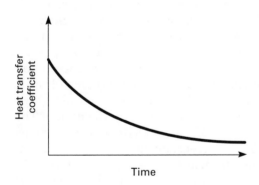

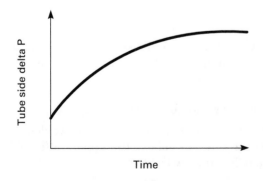

Figure 19.5 Effect of terminal-tube velocity on exchanger performance.

This is an exchanger that was designed to operate with a low tube-side velocity. Note how there is a rapid loss in the heat-transfer coefficient, as the tubes foul and plug, as a result of low velocity. The loss in heat-transfer coefficient U stops only when the terminal-tube velocity is reached in the unplugged tubes.

Note, also, how there is a rapid increase in the pressure drop in the tube bundle, as the tubes foul and plug, because of low velocity. The increase in ΔP stops only when the terminal-tube velocity is reached in the unplugged tubes.

It would seem that it might be best to design exchangers for initially high velocities and high pressure drops. It would seem that in practice, if we were to design exchangers for a low ΔP, after a few months, fouling would cause a high ΔP, anyway. This line of reasoning is valid for both the tube and shell sides of a heat exchanger.

Heat-Transfer Efficiency

"No fouling—no fooling"

This is the title of an important article[1] published many years ago. I had the honor of meeting the author, several years my senior, in 1965. It could be that his sort of plain-talking, hands-on applied science is a vanishing craft. I hope not. The substance of this famous article was simple. When designing heat exchangers, use the available pressure drop to maintain high velocities through the heat-exchanger equipment. Do not use safety factors in allowing for future pressure drop due to fouling. Use of such safety factors will force the mechanical engineer who designs the exchanger to use lower velocities in the design. Then the feared fouling will occur, and the ΔP safety factor will be consumed.

But suppose we are operating a heat exchanger subject to rapid rates of initial fouling. The start-of-run heat-transfer coefficient U is 120 Btu/[(h)(ft^2(°F)]. Four months later, the U value has lined out at 38. The calculated clean tube-side velocity is $1\frac{1}{2}$ ft/s. This is too low, but what can be done?

Multipass exchangers

It is possible to convert the two-pass tube bundle shown in Fig. 19.1 to the four-pass tube bundle shown in Fig. 19.6. This conversion is effected as follows:

- The center channel head pass partition baffle is cut out.
- Two off-center channel head pass partition baffles are welded in place, so that 25 percent of the tubes are above the upper baffle and 25 percent of the tubes are below the lower baffle.

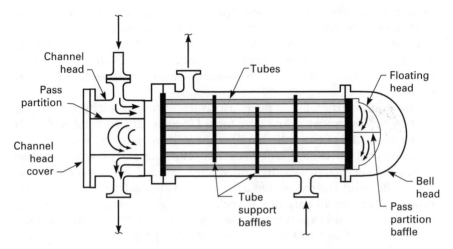

Figure 19.6 Four-pass tube bundle.

■ Both the channel head cover and the channel head tubesheet (see Fig. 19.1) must be remachined to accommodate the new baffles.

■ A new, center, pass partition baffle is welded in the floating head. The floating-head tubesheet must also be remachined.

The resulting four-pass tube bundle will have a tube-side velocity twice as high as it did when it was a two-pass exchanger: 3 ft/s. Experience has shown that in many services, doubling this velocity may reduce fouling rates by an order of magnitude. That is fine. But what about pressure drop?

When we convert a tube bundle from two to four passes, the pressure drop increases by a factor of 8. For example, assume that the two-pass ΔP was 5 psig. With the same flow, the four-pass ΔP would be 40 psig. Let me explain:

■ Pressure drop increases with the square of velocity.

■ If the velocity doubles, the pressure drop would go up by a factor of 4.

■ Pressure drop increases directly (linearly) with the length that the flow traverses.

■ The fluid must go twice as far in a four-pass as in a two-pass tube bundle.

■ Four times 2 equals 8.

Quite likely, even after several years of operation, the pressure drop of a four-pass exchanger, will be greater than the ΔP of a two-pass exchanger, in the same service. Quite likely, the initial U value of a four-pass exchanger, days after it has been returned to service, will be only slightly higher than the U value of a two-pass exchanger. However, the four-pass exchanger will maintain its U value with time, far better than will the lower-velocity two-pass tube bundle heat exchanger.

The eightfold increase in pressure drop is certainly a stiff price to pay for this improvement in the long-term U value. But remember this. It is the clean ΔP of the tube bundle that will increase by a factor of 8. Let's say that the tube side of an exchanger is currently operating, after 2 years of service, in a badly fouled state. Its ΔP, as measured in the field, is 20 psig. The calculated ΔP for clean tubes is 5 psig in a two-pass configuration, or 40 psig in a four-pass configuration. After 2 years of operation, I would expect the exchanger in the four-pass configuration, to have a ΔP of 50 to 60 psig, rather than 160 psig.

Why? Well, because the doubling of the tube-side velocity has promoted turbulence, which retards the accumulation of fouling deposits. Of course, this is exactly the reason why we exercise. The increased flow of blood through our arteries prevents plaque from sticking to the walls of our blood vessels. The plaque, a fatty deposit derived from cholesterol, restricts the flow of blood, which causes high blood pressure and eventually strokes.

Shell side vs. tube side

We can have any even number of tube-side passes: two, four, six, eight, etc. But this certainly limits our flexibility to optimize the tube-side velocity.

The shell-side cross-flow velocity may be altered in much smaller increments, by changing the tube support baffle spacing. This is one advantage of placing the fluid with the poorer heat-transfer properties on the shell side. But there is another, far more critical, advantage in placing the fluid with the poorer heat transfer properties on the shell side.

Laminar flow.   Think about water flowing slowly in a channel. Will the water in the center of the channel flow faster or slower than the water along the sides of the channel? Experience teaches that water in the center of a channel will flow faster than along the sides of the channel. Moreover, if the water flow is really slow, the water creeping along the side of the channel will barely mix with the bulk of the water flowing in the center of the channel. This is called *laminar flow.*

Fluids in laminar flow transfer heat very poorly. The slow-moving fluid creeping along the heat-transfer surface does not particularly mix with the bulk of the fluid. While the slow-moving fluid may get hot, it does not transfer its heat very efficiently to the flow of the bulk of the fluid. This is called *film resistance.*

Laminar flow is very bad for heat transfer. After fouling, it is the second biggest reason for low U values. Laminar flow is caused by two factors:

- Low velocities
- High viscosities

A low velocity for liquids is <2 or <3 ft/s. Velocities of >10 or >12 ft/s may cause erosion of metal surfaces and should be avoided.

A low viscosity is <2 or <3 centipose (cP) or centistokes (cSt). Tap water has a viscosity of about 1 cP. A viscosity of ~50 cP is quite high. Warm maple syrup has this sort of viscosity. The viscosity of vapors is almost always very low.

When we cool a liquid off, its viscosity markedly increases. I cannot generalize—it depends on the fluid. But I can say that increasing the viscosity of a fluid from 2 to 40 cP could reduce the observed heat-transfer efficiency (U) from 100 to 25. I know this from my experience in preheating cold, viscous, Venezuelan crude oil, off-loaded from tankers.

The best way to diminish the effect of laminar flow is to place the higher-viscosity fluid on the shell side. The shell side of an exchanger is far more resistant to heat transfer loss due to film resistance than is the tube side because of

- The vortex shedding shown in Fig. 19.2
- The rapid changes in direction, due to the tube support baffles

Purely in terms of heat transfer, it follows that the higher-viscosity fluid should be placed on the shell side. Sometimes, pressure and corrosion force the designer to allocate the higher-viscosity fluid to the tube side. Also, your maintenance department would vastly prefer that the fouling fluid be placed on the tube side.

Exchanger Cleaning

To disassemble an exchanger may require a lot of work. However, if only the tube side has to be cleaned, the amount of labor required may be halved. Referring to Fig. 19.6:

- The channel head cover is removed.
- The bell head (or floating-head cover) is dropped.
- The floating head is removed.

Each individual tube can now be cleaned by one person with a long lance and a hydroblast machine.

On the other hand, cleaning the shell side requires pulling the entire bundle out of the shell. This is a massive undertaking requiring a special bundle pulling machine and a large crew of pipe fitters. Then comes the hard part, cleaning the bundle.

If the deposits are soluble in a solvent, the shell may be soaked in a chemical bath. Sometimes this bath must be kept hot to dissolve the deposits, at a reasonable rate. Sometimes, the deposits must be attacked with a hydroblast machine. If the design engineer has selected a correct *tube pitch,* cleaning the shell side of an exchanger with a hydroblast machine may be a reasonable proposition.

Mechanical Design for Good Heat Transfer

Selecting proper tube pitch

From a theoretical heat-exchanger perspective, the triangular tube pitch shown in Fig. 19.7 is best. The term *pitch* refers to the geometry and distance between the holes drilled into the tube support baffles. A *triangular pitch* means the holes are drilled in the pattern of a 60° equilateral triangle. For example, we might have a ¾-in tube (OD) on a 1-in triangular pattern. The advantages of this tube pitch are

- Many tubes are squeezed into a small shell ID.
- The tube spacing promotes turbulence and hence good heat transfer.

This all results in reduced capital cost. The disadvantage of this tube pitch is that, except for the outer few rows of tubes, the shell side cannot be cleaned by hydroblasting. Obviously, this sort of arrangement should never be used in a fouling service, unless the deposits can be chemically removed.

The second example shown in Fig. 19.7 is for ¾-in tubes, also set on a 1-in spacing. But this time, the arrangement is a square pattern. This pattern reduces the number of tubes in a shell of a fixed ID. Also, less turbulence is created by the tubes. Some fluid may flow between the tubes, without encountering the tubes at all. All this reduces heat-transfer efficiency. But hydroblasting between the tubes is usually possible—depending on the extent of tube bending and twisting.

The third example shown in Fig. 19.7 is my favorite for sensible-heat transfer in fouling service. This consists of 1-in tubes set on 1½-in rotated square pitch. The pitch layout is the same as that of the square pitch. It is just the tube bundle that is rotated by 45°. The shell-side fluid cannot flow without interference between the tubes. Hence, the tubes promote turbulence, and improve heat transfer. The large tubes,

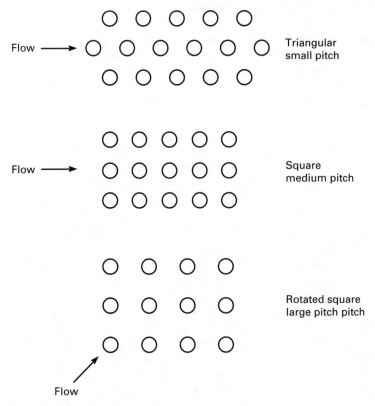

Figure 19.7   Square, large-pitch tubes are easier to clean.

and especially the doubling of the space between the tubes, reduce the heat-transfer surface area that can fit into a shell of a given ID. But the primary objective of the 1-in tubes, on a $1\frac{1}{2}$-in rotated square pitch, is that it is relatively easy to hydroblast.

Two-pass shell

Most of the heat exchangers in your plant are designed with the shell-side inlet and shell-side outlet at opposite ends of the shell, as shown in Fig. 19.6. However, you may have noted a few exchangers in which both the shell-side inlet and shell-side outlet are next to the channel head. This is a two-pass shell-side exchanger, of the type shown in Fig. 19.8.

The critical feature of this exchanger is the *longitudinal baffle,* which extends down the length of the shell. The baffle is fitted directly to, and thus becomes physically part of, the tube bundle. The function of the baffle is to force the shell-side fluid to flow down the entire length of the shell and back again to the shell-side outlet nozzle. Such an exchanger is truly countercurrent. The engineer does not need to apply the F factor derating coefficient to the log-mean temperature driving force, as would normally be used for nontrue countercurrent flow.

This would make it seem that the two-pass shell configuration is inherently more efficient than the ordinary single-pass shell exchanger. And it is, in theory. Then why do we see so few of these exchangers in service?

The reason illustrates the true nature of the shell-and-tube heat exchanger. It is a compromise between an ideal heat-transfer configuration and practical mechanical limitations. In this case, the difficulty is preventing leakage around the longitudinal baffle. Such leaks permit the shell-side fluid to short-circuit the tube bundle; that is, a percentage of the inlet flow, may flow directly to the outlet nozzle. In extremely serious cases, I have seen the bell head (shown in Fig. 19.8) 100°F colder than the shell-side outlet temperature.

When new, the longitudinal baffle seal, which is composed of a number of foil strips, extends along the length of the shell. These strips press up against the ID of the shell, and effectively seal off the upper half of the shell from the lower half, as illustrated by Fig. 19.9. Unfortunately, the first time the shell is pulled for maintenance, these foil strips are wrecked. Also, unless great care is taken by maintenance personnel when installing the new or replacement longitudinal baffle sealing strips, the heat-transfer efficiency of the exchanger will be greatly reduced after it is returned to service. As a result of this deficiency, most operating companies severely restrict the use of two-pass shell-side heat exchangers.

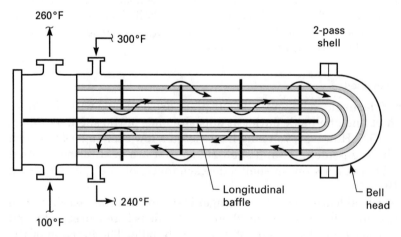

Figure 19.8 Two-pass shell exchanger.

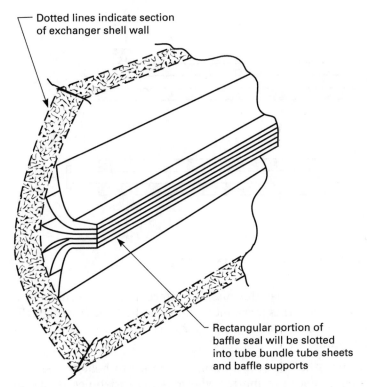

Dotted lines indicate section
of exchanger shell wall

Rectangular portion of
baffle seal will be slotted
into tube bundle tube sheets
and baffle supports

Figure 19.9 Detail illustrating how a modern longitudinal baffle seals up against the inside of the exchanger shell.

Double-pipe exchangers

Also called "hairpin" exchangers, the double-pipe heat exchanger, as shown in Fig. 19.10, is the simplest of all types we have discussed. It is nothing more than two concentric pipes. The inner pipe might be of 2 in diameter; and the outer pipe, 3 in. Double-pipe exchangers are true countercurrent-flow heat exchangers. They are rugged and are reasonably cleanable. Moreover, they are not subject to bypassing. The only problem with them is cost per square foot of heat-transfer surface area. A single pipe or tube simply cannot have much area.

Exchangers of less than a few hundred square feet can be economically designed in a hairpin configuration. Larger exchangers are, for economic factors, shell-and-tube types.

Fin tubes

You may have seen bundles constructed with serrated, or very small, fins covering the exterior of the tubes. These are called *low-fin-tube*

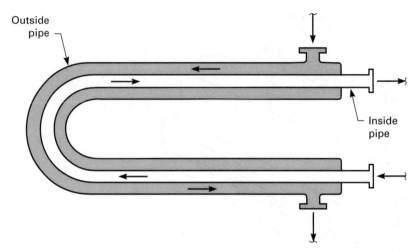

Figure 19.10   A double-pipe countercurrent exchanger.

bundles. These fins increase the outside surface area of the tubes by a factor of 2.5. However, this does not mean that the heat-transfer efficiency of the exchanger will increase by 250 percent. Two factors curtail this improvement:

1. Let's assume that the controlling resistance to heat transfer is inside the tube. It does not matter whether this resistance is due to fouling, high viscosity, or low flow. The heat flow must still pass through the smaller surface area inside the tube. The smaller area acts as a choke point that restricts heat flow. Let's assume that 90 percent of the resistance to heat transfer is inside, and 10 percent outside, the tubes. The overall increase in the heat-transfer capacity of the exchanger would, at best, be roughly 15 percent.

2. Actually, retrofitting a tube bundle with low fin tubes often reduces heat-transfer capacity. This happens when the controlling resistance to heat transfer is shell-side fouling. The fouling deposits get trapped between the tiny fins. This acts as an insulator between the shell-side fluid and the surface of the tubes. In severe shell-side fouling services, I have replaced fin tubes with bare tubes, and doubled the heat-transfer duty on the exchanger.

In summary, low fin tubes may be used to advantage only when the controlling resistance to heat transfer is the shell-side fluid itself.

Heat-transfer resistance

The resistances of electric resistors arranged in series may be added to obtain the overall circuit resistance. This concept also applies for a heat exchanger. The following five resistances to heat transfer are added together to obtain the total resistance to heat transfer:

- The tube-side fluid
- The fouling deposit inside the tubes
- The metal wall of the tube
- The fouling deposit around the outside of the tube
- The shell-side fluid

The sum of these five factors is called the *overall resistance* to heat transfer. The reciprocal of the overall resistance is termed U, the overall heat-transfer coefficient.

Reference

1. G. H. Gilmore, "No-Fouling—No Fooling," *Chemical Engineering Progress,* July 1965, vol. 61, no. 7, pp. 50–56.

20

Fired Heaters: Fire- and Flue-Gas Side

Draft and Afterburn; Optimizing Excess Air

Next to the efficiency of the trays in a distillation tower, the efficiency of a fired heater is the most critical factor in saving and/or making money for the process plant. The primary objectives in operation of a fired heater are to

- Keep the fire in the firebox
- Avoid excessive heat density in the firebox
- Maximize the process heat absorption for a given amount of fuel

These objectives are equally important in the operation of fired boilers whose principles of operation on the fire side and flue-gas side are essentially the same as those of process plant fired heaters.

The two governing modes of heat transfer in the construction of a fired heater are radiation and convection.

In the firebox, heat transfer by radiation reigns supreme governed by Lambert's laws, as follows:

$$Q_R = A \cdot \varepsilon (T_{\text{fb}}^4 - T_{\text{tm}}^4)$$

where Q_R = radiant heat-transfer rate, Btu/h

A = surface area of the radiant tubes in the firebox, ft^2

ε = emissivity factor

T_{fb} = temperature, °F, of the radiant surface, which is essentially the firebox temperature. The reason for this is that the flames heat not so much the tubes as the refractory, and the refractory then reradiates the heat to the tubes, so the main heat source becomes the refractory.

T_{tm} = the receiving metal temperature, °F (this is the radiant tube metal or tube skin temperature).

The amount of heat transferred to the process fluid in the convective section of a heater is governed by

$$Q_c = A \cdot U(T_{fg} - T_{bulk})$$

where Q_c = convective heat-transfer rate, Btu/h

A = surface area of the tubes, ft^2

U = a heat-transfer coefficient, Btu/[(h)(ft^2)(°F)]

T_{fg} = temperature of the flue gas, °F

T_{bulk} = temperature of the process fluid flowing inside the tubes, °F

Note that the temperature driving force for radiant heat transfer $(T_{fb}^4 - T_{tm}^4)$ is always a very large number as compared to the temperature driving force for convective heat transfer $(T_{fg} - T_{bulk})$. For this reason, in nonfouling services, we used finned tubes in the convective section of our fired heaters to increase the surface area of the tubes. The tubes in the firebox or radiant section are made of high-chrome steel, capable of withstanding firebox temperatures of up to 1100 to 2000°F depending on the severity of operation for which the heater has been designed. (Some very high-severity heaters may be designed for firebox temperatures of >2000°F.) Note especially that radiant section tubes are protected from overheating by the cooling effect of the process fluid flowing inside them, where "flowing" is the key word. The bare, unfinned, shock tubes are also cooled by the process fluid flowing inside them.

However, the convective-section finned tubes are not intended or designed to withstand such high temperatures, and so the tubes themselves are often made from low-temperature-rated carbon steel, whereas the fins, which are not much cooled by the process flow, are often made from low-chrome steel. As it is much easier to make finned tubes from just one type of metal instead of two, the furnace manufacturers will often choose to make the finned convective-section tubes entirely out of low-chrome steel in which case one could expect them to withstand temperatures of ≤1300°F. It is advisable to check what types of tubes you have in your furnace and *know* what temperature tolerances they have. The only thing that prevents the tubes in the convective section from overheating up to firebox temperatures is that we keep the fire in the firebox, and do not allow the fire to get up into the convective section. This may sound obvious, but it is surprising how many people seem to forget or ignore this fact.

A typical natural-draft gas–fired process heater is shown in Fig. 20.1. Suppose we gradually close either the stack damper or the air register; the flow of air into the firebox will then be reduced. If both the process-side flow and the fuel-gas rate are held constant, the following sequence of events occurs:

1. The heater process outlet temperature begins to increase as the excess air is reduced. This is because more heat is given to the process fluid and less heat goes up the stack.

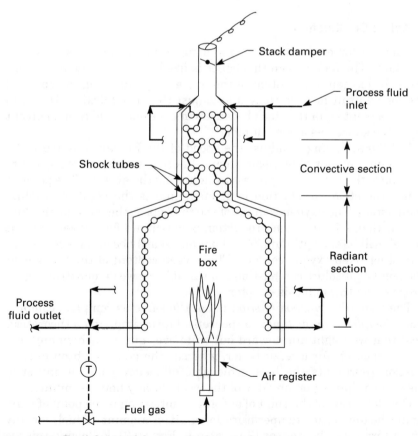

Figure 20.1 Typical natural draft gas-fired process heater.

2. The heater process outlet temperature declines as airflow is reduced past the *point of absolute combustion*. In this situation we have products of incomplete or partial combustion such as aldehydes, ketones, and carbon monoxide going up the stack. This sets the heater up for *afterburn* in the stack, and the heating value of the fuel is also effectively reduced.

Allowing a fired heater, boiler, or furnace to operate with insufficient air is hazardous because

- The products of incomplete combustion are hot and just waiting to catch fire and will ignite as soon as they find sufficient oxygen. This usually results in afterburn in the convective section or stack and can even lead to explosive detonations.
- The products of partial combustion are atmospheric pollutants.
- It is not possible to operate on automatic temperature control on the wrong side of the point of absolute combustion.

Absolute Combustion

Consider a forced-draft boiler producing 600-psig steam as shown in Fig. 20.2. The fuel rate on this boiler is fixed and we are going to optimize the oxygen (O_2) content of the flue gas by adjusting the speed of the forced-draft fan. Do we simply adjust the forced-draft (FD) fan to give 2 percent O_2 in the stack because someone once said that 2 percent O_2 in the stack was a good number?

No! We are going to adjust the speed of the FD fan to produce the maximum amount of 600-psig steam. In other words, we are going to maximize the heat to the process by adjusting the air rate. The point at which the steam production is at maximum is the point of absolute combustion. The oxygen content of the flue gas at the point of absolute combustion, where steam production is maximum for a given amount of fuel, will be the optimum; it will not necessarily be at 2 percent O_2 or any other fixed oxygen content. The oxygen content of the flue gas at the point of absolute combustion is a variable, as we demonstrate while progressing through this chapter.

The term *absolute combustion* is not the same as complete combustion. *Complete combustion* is a theoretical term, implying a theoretical goal that we might aim toward but will never quite reach on any real process heater, furnace, or boiler, whereas the point of absolute combustion represents the best achievable efficiency point of any such piece of equipment on any day of the week, at any hour or minute.

One definition of the point of absolute combustion is the point of maximum heater outlet temperature for a given amount of fuel, for any given furnace or heater (as illustrated in Fig. 20.3). Following this we

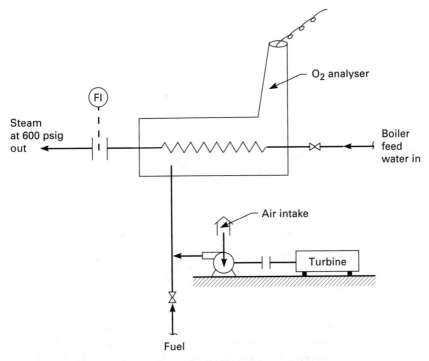

Figure 20.2   Forced draft–fired boiler; simplified schematic drawing.

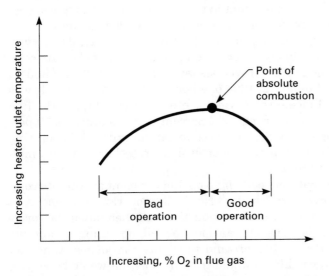

Figure 20.3   Point of absolute combustion in terms of heater outlet temperature.

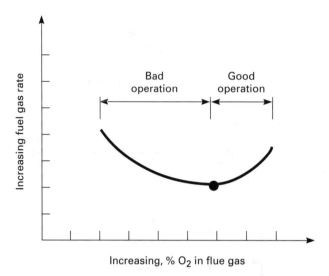

Figure 20.4 Point of absolute combustion in terms of maximum achievable combustion of fuel.

can say that the point of absolute combustion is also the point of best or maximum achievable combustion for a given amount of fuel for a given furnace, heater, or boiler (as illustrated by Fig. 20.4). Also please note that the point of absolute combustion in terms of either maximum heater outlet temperature or minimum fuel fired for a given outlet temperature would correspond to the same flue-gas oxygen content for a given furnace at the same moment in time. However, we must also note that this flue-gas oxygen content that corresponded to the point of absolute combustion on, say, heater 3304-A on Tuesday, January 2, 1996 at midday is not likely to correspond to the point of absolute combustion on the same heater on Wednesday, January 3, 1996 at midday, or even Tuesday, January 2, 1996 one hour or one minute after it was first determined. Indeed, the oxygen content (percent O_2), of the flue gas typically varies in the range of 2 percent O_2 to as much as 10 percent O_2 depending on the design, mechanical integrity, and operating characteristics of the equipment, such as firing rate, burner type, ambient weather conditions, etc.

Did the oxygen analyzer really help us find the point of absolute combustion for the forced-draft boiler? No, it did not. However, once you have found the point of absolute combustion and then noted the corresponding flue-gas percent O_2, as long as all operating conditions remain constant for the heater (and in reality the longest time you could hope that *all* conditions would truly be constant would be ~½ h), then you could use the percent O_2 in the flue gas as a rather secondary guideline for that short period of time until the operation conditions changed.

Oxygen starvation

If you try to operate a furnace, fired heater, or boiler with too little combustion air to starve the burners of oxygen to "smother" or "bog down" the firebox, then you will likely cause "afterburn" or secondary combustion in the stack, you will not be able to operate on automatic temperature control, and may even destroy the equipment altogether.

This type of operation can also be described as operating on the wrong side of the point of absolute combustion. Take another look at Figs. 20.3 and 20.4; we are talking here about the portion of the curves marked as "bad operation." What would be the color of the flue gas as it emerged from the stack during this type of operation? Would it be black?

No, it would not necessarily be black.

If we were burning heavy industrial fuel oil with a very high carbon:hydrogen ratio, then yes, very soon after we began to operate on the wrong side of the point of absolute combustion, the stack gases would turn black. After all, the black color is merely soot or unburned carbon. However, if we were burning pure hydrogen, the stack gases would never turn black no matter how much we starved the firebox of oxygen. If we were burning natural gas, then as we crossed over to the wrong side of the point of absolute combustion, we might initially just be able to see the stack gases turn only a pale yellow, then if we continued to decrease the combustion-air rate, the stack gas color might progress from pale yellow, to dark yellow, to light orange, to dark orange, to brown, and finally to black. The color of the stack gases is not an accurate indication of whether we are using enough combustion air, as the point at which the stack gases finally turn black depends on the carbon:hydrogen ratio of the fuel. Also, while continuing to fall below the point of absolute combustion, we accelerate the production of atmospheric pollutants such as aldehydes, ketones, light alcohols, carbon monoxide, and other products of the partial combustion of light hydrocarbons in the stack gases.

If we have the furnace on automatic temperature control while we are not using enough combustion air, and if the control valve on the fuel gas then opens to allow more fuel to the burners in order to either maintain or perhaps increase the furnace outlet temperature, the extra fuel will not burn efficiently. In fact, the extra fuel is likely to *reduce* the heater or furnace outlet temperature rather than increase or maintain it, as there is already a shortage of air and it cannot burn properly and tends to cool the firebox. The automatic temperature controller then senses the reduction in heater outlet temperature and takes action by further increasing the fuel rate. Thus the furnace will spiral into an increasingly dangerous condition as the outlet temperature continues to fall and the furnace is left on automatic control. The way out of this situation is to put the furnace on manual control, and manually reduce the fuel gas back to the point of absolute combustion, by looking for the maximum heater outlet temperature for a given amount of fuel.

However, if we operate on manual without enough combustion air, we will share the experience explained by one oil refinery operator, as follows:

> Our crude unit heater is very old; we have to operate it manually. We have three air blowers to produce the combustion air; we always run two with one on standby. Once a week we switch the blowers by shutting one down and turning on the one that was idle.
>
> One week I was switching the blowers when I forgot to shut one of the two down before starting up the third blower. I then had all three blowers running at once. Our normal heater outlet temperature target is 680°F, but while I had all three blowers on, the outlet temperature rose rapidly to 740°F.

Rather than cutting the fuel gas back to reach a 680°F outlet temperature using more air, and to save fuel, our friend went on to explain how, in peril of exceeding a management directive to maintain a certain oxygen content in the stack gases, he in fact hastily shut down the third blower and therefore dropped the furnace temperature back to 680°F but also obviously continued to run fuel-rich, without enough air.

Which is cheaper, air or fuel? Of course, the answer is *air*. Are we trying to make the fuel gas to some particular oxygen-content specification, or are we trying to save fuel? Certainly the answer is that we want to get the most heat we can out of every pound of fuel. Well, then, dear reader, let us neither set nor encourage the use of these stack or flue-gas oxygen targets; they are meaningless and often cause more harm than good.

Appearance of the firebox and flames

Regardless of and in preference to oxygen analyzer results, we must consider the appearance of the firebox and flames when we assess whether more combustion air is needed:

If the firebox appears bright and clear, then there is more than enough oxygen (a crystal-clear firebox has a large excess of oxygen).

If the firebox appears hazy and the flames are long, licking, yellow, and smoky-looking, then there is not enough oxygen.

To be just right, the firebox should have a very slight haze and all the flames should be compact and not searching around the firebox looking for oxygen.

Do not stand directly in front of an inspection port while you open it, in case there is positive pressure behind it.

Flame color depends on fuel composition. Gas often burns blue, but heavy fuel oil burns yellow. A yellow flame is caused by thermal cracking of the fuel. There is nothing wrong with a yellow flame; it is the general shape of the flame which is important. If in doubt as to the right flame shape for a particular furnace, contact the burner manufacturer for details.

The temperature of the inside of a firebox can also be estimated by its visual appearance. Table 20.1 provides a color/heat guide.

Secondary combustion or afterburn

In the preceding paragraphs we discussed how operation without enough combustion air, on the wrong side of the point of absolute combustion, leads to incomplete and therefore inefficient combustion of fuel. Apart from the fuel wastage, there is another important disturbing and frequently occurring problem associated with this bad mode of operation. The products of incomplete combustion of the fuel, and in more severe cases even the unburned fuel itself, flow with the flue gas up through the convective section and up the stack. A certain amount of air leaks in through the convective section and stack from outside the heater. When these hot, combustible hydrocarbons in the form of unburned fuel or products of incomplete combustion mix with the extra air that is leaking in from the environment, reignition is liable to occur. We call this *afterburn,* or *secondary ignition,* meaning there is now fire in the convective section, but the fire is supposed to be contained within the firebox.

When combustible materials or unburned fuel reignite in the convective section, a dramatic increase in flue-gas temperature will occur

TABLE 20.1 Visual Estimation of Temperatures

Color	Temperature, °F
Dark blood red or black red	990
Dark red	1050
Dark cherry	1250
Bright cherry	1375
Light red	1550
Orange	1650
Light orange	1725
Yellow	1825
Light yellow	1975
White	2200
Dazzling white	2730

from normal operation at, say, 700°F up to, say, 2000°F. The metallurgy of finned carbon steel convective-section tubes is not designed to withstand such temperatures. The fins will become oxidized and, when cool, may become brittle and thicker than before, thus restricting flue-gas flow. The convective-section tubes themselves will become warped and bent and thus restrict flue-gas flow still further.

Just such a thing happened as a result of another well-intentioned but misguided attempt improve fuel firing efficiency by reducing the oxygen content of flue gas in a fired heater. This heater, which was also manually operated, had an oxygen analyzer probe installed in the convective section. The operators were given a target range to run between 2 and 6 percent O_2 (note that this range would in many cases actually include the point of absolute combustion).

The inside operator (operator A), noting that the stack gases had 8 percent O_2, began to cut back on the combustion air. As operator A pinched back on the air register, the convective-section oxygen dropped slowly from 8 to 7.5 percent O_2. The outside operator (operator B) went to check on the firebox and, seeing that it was full of long, licking, yellow, smoky flames and had taken on a very hazy appearance, reported back to operator A that he considered the amount of air in the firebox was insufficient.

Meanwhile operator A looked again at his O_2 analyzer output, which showed 7.5 percent O_2. He told operator B that there was plenty of oxygen because the analyzer showed 7.5 percent O_2, so he must be mistaken and, in fact, that he needed to go and pinch back still further on the air register to cut back the combustion air.

After cutting back on the air registers, operator A noticed that the heater outlet temperature dropped, so he raised the fuel rate to maintain the heater outlet temperature. He did not realize that this sequence of events was an indication of insufficient combustion air, because the convective-section oxygen analyzer still showed 7 percent O_2, so he asked operator B to pinch back still further on the air registers.

At this point black smoke started coming out of the stack, and several of operator B's colleagues began to suggest that they should start to open up the air registers and increase the combustion air and not reduce it any further. However, operator A, now reading 6.5 percent O_2 in the convective section on his O_2 analyzer, was determined to reach his 6 percent O_2 target , and requested a further cutback on the combustion air.

When the combustion air was cut back this time, fire started to come out of the stack. Now Operator A was forced to admit that they would have to increase the combustion air again. The problem they then faced was that the amount of draft in the heater seemed to be less than before and they were not quite able to reestablish the same airflow. The reason for the restricted airflow was that they had caused afterburn in the convective section and damaged the convective-section tubes, which now restricted the flow of flue gas and hence the flow of combustion air to the burners.

Using the concept of absolute combustion in operations

The only correct oxygen target is that firebox oxygen content which maximizes the process-side heat absorption for a given amount of fuel, or if you prefer, we could say it is that the firebox oxygen target which minimizes fuel consumption for a given process-side heat absorption.

This correct percent O_2 in the firebox can be determined only by experimenting with the fired heater, furnace, or boiler in the field. This firebox O_2 will also vary for any given piece of fired equipment depending on a variety of factors, including mechanical integrity and firing rate. In reality, it is not the firebox oxygen content that we are looking for so much as the correct combustion-air rate for proper operation. The search for the correct combustion-air rate on any particular day is known as *heat-proving the heater.* Experienced operators will increase and decrease the combustion air to find that particular rate that minimizes the fuel rate for their designed process heat absorption which they will typically monitor in the form of heater outlet temperature or steam flow rate. Naturally, experienced operators have been heat-proving heaters since long before oxygen analyzers were invented. In view of this, it seems that we do not actually need an oxygen analyzer to run a piece of fired equipment efficiently. This is correct.

The concept is not to run the fired equipment to a specific oxygen target because the oxygen content (wherever measured) corresponding to the most efficient operation varies, but to aim instead to maximize heat absorption to the process stream for a given fuel rate, or to minimize fuel firing rate for the desired heat absorption by the process stream. Once we have heat-proved the heater and found this point of maximum heat absorption, or minimum fuel rate, which we call the *point of absolute combustion,* we will then need to operate the furnace with somewhat more air. We call this operating on the "good" side of the point of absolute combustion, as shown in Figs. 20.3 and 20.4.

If we do not ensure that we operate with some extra air to put us on the "good" side of the point of absolute combustion, then we run the risk of getting into oxygen starvation on the wrong side or "bad" side of the point of absolute combustion.

Automatic operation linked to process outlet temperature while on the bad side of absolute combustion is potentially hazardous because the heater outlet temperature will drop as a result of the reduced heating efficiency of the fuel. The automatic control will then call for more fuel, which, in turn, produces more oxygen starvation. The only way out of this situation is to put the furnace back to manual operation and manually increase the combustion air or reduce the fuel to get back to the good side of the point of absolute combustion.

Note that combustion air is relatively inexpensive compared to fuel; it is irrational to cut back on combustion air and waste fuel by inefficient burning. Also note that to operate fired equipment efficiently, we need to heat-prove it, or test to find the point of absolute combustion, on a regular basis, preferably at least once a day.

Flue-gas oxygen and tramp air

Suppose we have a natural-draft heater operating very efficiently on the good side of the point of absolute combustion. The oxygen content of the firebox gases (just below the shock tubes) is 2.5 percent oxygen as shown in Fig. 20.5. What do you think the oxygen content of the flue gases in the stack will be?

The answer is it will *always* be more than the firebox oxygen content, because there are *always* air leaks in the convective section and stack that, with the slightest negative pressure in the stack or convective section, will allow air to leak in. These air leaks are collectively known as "tramp air."

The amount of tramp air will vary according to

- The mechanical integrity of the heater, i.e., the number of holes in the casing
- The firing rate and draft balance of the heater

The oxygen measured in the stack is the sum of the unused oxygen from the firebox plus the oxygen from tramp air, drawn into the convective section and stack.

We have already shown that it is wrong to operate a furnace to an arbitrary oxygen target. It is entirely incorrect to adjust the furnace combustion air based on stack gas and/or convective-section flue-gas oxygen readings, because—as a result of the inevitable air in-leakage—the stack gas and/or convective-section flue-gas oxygen content in no way represents the amount of oxygen in the firebox. It is also impossible to offset the stack or convective-section oxygen readings to represent the firebox oxygen content because the amount of air in-leakage varies so much with time as conditions change.

Unfortunately, many heaters have oxygen sensors in the stack and not in the firebox. An oxygen sensor in the firebox may provide a useful guide, as we mentioned earlier, although it is not actually needed to run the heater efficiently. However, an oxygen analyzer in the stack or

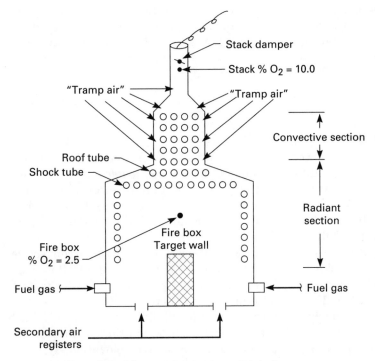

Figure 20.5 Natural-draft heater showing tramp air leaks.

convective section is really useful only in operations if there is also an oxygen analyzer in the firebox; in such instances it could be used to trend air leaks across the convective section. However, we must never ever use an on-line analyzer for technical analysis or test work. Oxygen-content field measurements should always be taken with a single portable oxygen analyzer, such as that made by Teledyne. We must also be careful to calibrate the portable instrument for the range in which we want to use it. Cylinders of test gases with specific oxygen contents are available for calibration purposes (ask your plant instrument technicians if you can borrow theirs).

In view of the above, it is clear that if the combustion of airflow is being adjusted on the basis of the oxygen content of the stack or convective section gases (i.e., from readings taken on an oxygen analyzer with probe located in the stack or convective section), it is very likely that there will be afterburn or secondary ignition in the stack or convective section. This scenario was discussed earlier in this chapter in the section on secondary combustion and afterburn. After all, the fire is supposed to be contained within the firebox, and not in the convective section or stack; therefore, if we intend to install an oxygen analyzer, the analyzer should be monitoring oxygen in the firebox itself. Lone oxygen analyzers monitoring oxygen in the convective section or stack are so misleading as a result of the tramp air, that they are a positive hazard to efficient and safe operation.

Draft

Draft readings

Draft readings are a comparison of two pressures taken *at the same elevation* and are traditionally quoted in *inches of water gauge*. Whereas it is true that if you open an inspection port on a heater in an area of positive draft, you are liable to singe your eyebrows; if you wish to really study or make sense of a set of draft readings for a furnace, you must first normalize the data.

In Fig. 20.6, we see a simple natural-draft heater with no convective-section tubes. The laws of hydraulics tell us that fluids flow from regions of high pressure to regions of lower pressure, and yet the draft readings in Fig. 20.6 seem to contradict this principle.

Notice that the draft readings are made at different elevations. Each measurement is in reality a comparison between the densities of the gas both inside and outside the furnace at a given elevation. The temperature difference is the main reason for difference in density inside and outside the furnace because the molecular weights of the furnace gases and of air are approximately the same.

To make sense of the draft measurements so that we can use them to evaluate furnace pressure drop, proceed as follows:

1. Make a datum line across the top of the stack as shown in Fig. 20.6.

2. For each draft reading, *add* on the pressure exerted by the appropriate static head of air.

Given: 150 ft of air = 2.22 in H_2O (or inches of water gauge), in our example, which is for 60°F ambient air at sea level. Figure 20.6 illustrates the principle.

The standard instrument which is used to take draft readings is the magnehelic delta-pressure (ΔP) gauge. However, if you wish to carry out a draft survey, it is possible to make some fairly good readings with nothing more complicated or expensive than a clear glass or plastic bottle, some water, and a length of clear plastic flexible tubing. Fill the bottle with water, insert the plastic tube into the neck of the bottle, and then attach the other open end of the plastic tube at the point where you wish to take the draft reading. The difference in the vertical height (measured in inches) between the level of water in the bottle and the

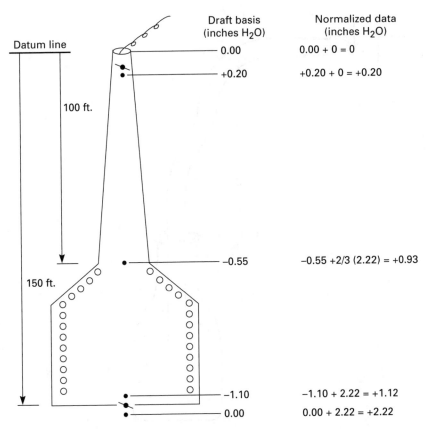

Figure 20.6 A simple natural-draft heater with no convective section tubes.

level of water drawn up in the tube will be your draft reading in inches of water gauge, as shown in Fig. 20.7.

Draft balancing

The air register and the stack damper are used together as a team to optimize the heater draft. Our aim in balancing the draft for a natural-draft or a balanced-draft heater is to maintain a small negative pressure of, say, −0.2 to −0.1 in of water, just below the shock tubes (please refer to Fig. 20.5), at the entrance to the convective section. At the same time we must maintain enough air to operate on the good side of the point of absolute combustion.

If we operate with a positive pressure in the firebox, although the burners may appear to operate normally, the hot flue gases will leak outward. This damages the roof arch supports and the steel structure so as to shorten the life of the heater.

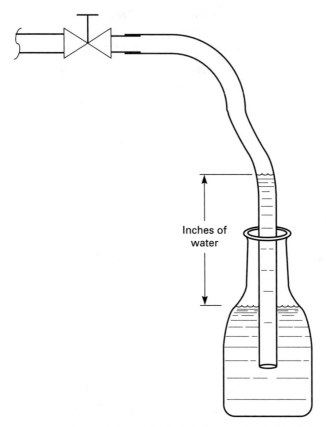

Inches of
water

Figure 20.7   A clear bottle and a flexible tube can be used to make approximate draft measurements.

As we close the stack damper, the pressure at the convective-section inlet will increase; that is, we will have less draft. This will then reduce the rate of flue gas to the convective section, minimizing the rate of air in-leakage to the convective section and the chances of afterburn occurring. However, if we close the stack damper too much, a positive pressure will develop at the convective-section inlet, which we must avoid.

Conversely, as we open the stack damper, the pressure at the convective-section inlet will decrease, so that we have more draft. If we have too much draft, we will increase the risk of afterburn as we are increasing the rate of air in-leakage in the convective section and stack.

We adjust the draft with the stack damper and maintain the combustion-air level by adjusting the air registers to accommodate the adjustments made on the stack damper.

Air Leakage

Evaluating fuel wastage due to air leaks

Convective-section air leaks reduce the thermal efficiency of the heater by mixing cold ambient air with hot flue gas. The resulting energy debit may be expressed as follows:

$$\Delta F = \frac{(T_s - T_a)\,(O_{2,s} - O_{2,c})}{500}$$

where ΔF = percent heater fuel wasted through lost-heat recovery in the convective section

T_s = stack temperature, °F
T_a = ambient temperature, °F
$O_{2,s}$ = percent oxygen O_2 in stack.
$O_{2,c}$ = percent oxygen in combustion zone, i.e., just below the shock tubes at the inlet to the convective section

Just for practice, let us suppose we have a stack temperature of 600°F, ambient air is 100°F, the convective section has 10 percent O_2, and the firebox has 6 percent O_2 measured just below the shock tubes. What percentage of fuel fired (ΔF in our equation) is being wasted by cold air leaking into the convective section?

$$\text{Answer:} \quad \Delta F = \frac{(10 - 6) \times (600 - 100)}{500} = 4\%$$

Minimizing fuel wastage due to air leaks

Following on from the preceding example, suppose we now cut back on the combustion air by closing the air registers a little. The oxygen content in the firebox is now 3 percent. The oxygen in the convective section will also be reduced to, say, 9 percent oxygen. Notice how the difference between the firebox oxygen and convective-section oxygen (which we refer to as ΔO_2) has increased. It is now 6 percent O_2. This is because the draft through the heater has increased (i.e., the pressure is more negative), and more air is being sucked in through the holes or air leaks in the convective section. If we still had a stack temperature of 600°F with ambient air at 100°F, we would now be wasting 6 percent of the fuel fired. This is not a pleasing result, as we can see extra fuel will have to be burned in the firebox to offset the increased air in-leakage.

Suppose we put the air registers back as they were and pinch on the stack damper instead. So let's suppose we restrict airflow via the stack damper until the firebox oxygen goes down from our base case value of 6 percent, to 3 percent. The oxygen in the convective section will also be reduced to, say, 5 percent. In this case, we see that the ΔO_2 is also much less, only 2 percent. This is because we now have less draft through the heater, i.e., the pressure in the convective section increased and so has reduced the air in-leakage rate. This also illustrates how the air leaks themselves vary in accordance with the combined operation of stack damper and air register.

Patching air leaks

To suppress afterburn and minimize energy losses caused by in-leakage of cold ambient air, any holes in the heater walls, convective section, ducts, etc. should be patched. Also make sure that inspection ports are closed. Leaks can be detected on stream to a certain extent by visual inspection (crumbling chalk dust or dropping a little baking powder past a suspected leak will pinpoint the leak).

During a turnaround, a smoke test should be carried out as follows:

- Close the stack damper.
- Ignite colored smoke bombs, or even old tires in the firebox.
- Turn on the forced-draft fan if there is one, to assist the smoke.
- Watch to see where the colored smoke emerges as these points are the sources of the air leaks.

Patch the leaks using heavy-duty aluminum tape, insulating mud, or silicone sealers, and weld up any loose sheet metal.

Efficient Air/Fuel Mixing

The function of a burner is to mix oxygen, in the form of air, with the fuel so that the fuel will burn more efficiently. Burners are available in a variety of different designs, all engineered with the intent to maximize air/fuel mixing efficiency, and in more recent times we have the added concern of also minimizing the formation of atmospheric pollutants. Please be aware that fuel will burn at the end of a pipe with no burner at all but the combustion will be far from efficient.

Some burners are fitted with primary- and secondary-air registers, as is the premix burner shown in Fig. 20.8. Air entering through the primary-air register mixes much more efficiently than does the air which enters through the secondary-air registers on such a burner. Thus we should maximize the use of primary air, and we do this by gradually opening up the primary-air register until the flame just begins to lift away from the burner tip. The remaining combustion air should be provided through the secondary-air register.

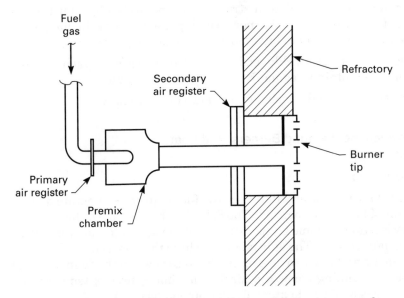

Figure 20.8 Schematic drawing of a premix burner showing primary and secondary air registers.

Optimizing Excess Air

The term *optimizing excess air* as we have seen, does not refer to operation at some arbitrary oxygen level; instead, consider the following in relation to your heater, and you will be optimizing excess air:

- Minimize fuel rate for a given heater outlet temperature (or steam production in a boiler), then operate at 0.5 to 1 percent higher oxygen.

- Maximize primary air to burners, where burners have primary and secondary air.

- Adjust draft to minimize air in-leakage while maintaining a small negative pressure at the entrance to the convective section.

- Close openings to pilot lights, sight-ports, and other holes around burners. (Combustion air only mixes properly through the burner air register.)

- When operating at reduced firing rates, shut down some burners if possible, as burners work more efficiently when operating close to their design capacity (also remember to close air registers on idle burners).

- Minimize poor lateral air distribution in the firebox by adjusting air registers on individual burners. Having a low airflow in one part of the heater will lead to higher overall oxygen requirements.
- Keep burners clean. Plugged burner tips increase oxygen requirements; maintain a regular burner cleaning program.
- Pay attention to the visual appearance of the firebox.

Air Preheating, Lighting Burners, and Heat Balancing

Air preheaters

A typical air preheater will reduce the fuel required to liberate a given amount of heat by 10 percent. The debit for this improved thermal efficiency is a higher flame temperature, and the possibility of overheating the radiant section. The only instance where there is a clear advantage to fit an air preheater to an existing furnace is when the firebox of that furnace is running below a desirable maximum firebox temperature. Three types of air preheaters are in common use:

- Direct heat exchange between flue gas and air
- Heat exchange via intermediate circulating oil
- Heat exchange via massive heat-transfer wheel, packed with metal baskets (Lungstrom type)

The wheel type, which is probably the most common in use, is subject to air leaks across the mechanical seals both internally between the air and flue-gas sections and also externally, to the mechanical seal. This leakage is identified by increased oxygen content in the flue gas, low flue-gas outlet temperature, and a greater temperature loss in the flue gas than rise in the air temperature across the preheater, as shown in Fig. 20.9.

In general, air preheater leaks are a problem because they

- Reduce the thermal efficiency of the air preheaters.
- May reduce combustion air so that it falls below that required for absolute combustion, resulting in destructive afterburn. (*Note:* air preheaters themselves are often destroyed by afterburn.)
- May cause the forced-draft fan to operate out of character, requiring higher driver horsepower.

All air preheaters are subject to corrosive attack caused by condensation of sulfur trioxide. At a 150-ppm sulfur in the fuel gas, operating

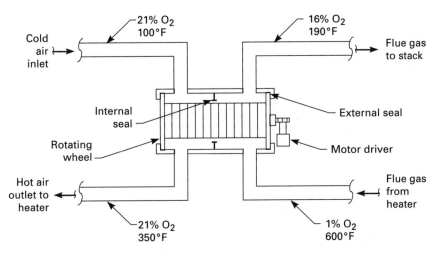

Figure 20.9 Rotating-wheel-type air preheater with leaking seals.

experience shows that a minimum temperature of 350 to 400°F minimizes corrosive attack. Uneven cooling in the preheater results in the
need to keep the flue-gas outlet temperature 50 to 100°F above the calculated SO_3 dew point. The flue-gas outlet temperature can be increased
by use of an air preheater air-side bypass, or by increasing the excess air
in the firebox (this second method is discussed further in the section on
heat balancing).

Heater thumping or vibration

Preheater vibration. Air preheaters or any type of waste-heat recovery
device designed for horizontal flow across vertical tubes, may be subject to vibration produced by the velocity of gas across the tube banks.
The velocity produces a vortex-shedding wave pattern that could correspond to the natural harmonic frequency of the tube bank. If the natural harmonic frequency is reached, excessive vibration of the tubes
will occur. Redesign of the internal baffle system by inserting dummy
baffles can stop the vibration.

Vibrations in fired equipment and afterburn. Balanced-draft or induced-
draft furnaces and boilers are intended to be operated with a small negative pressure (ca. −0.1 in H_2O), just below the first row of convective
tubes, i.e., just below the shock tubes. If we operate such a piece of
equipment with a severe shortage of air in the firebox and massive air

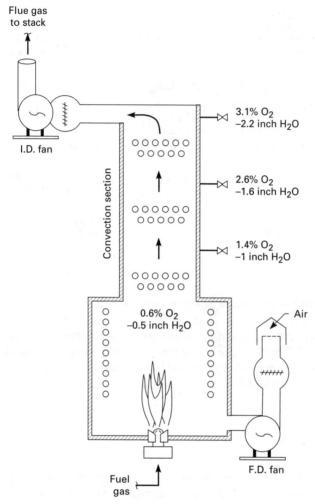

Flue gas
to stack

I.D. fan

Convection section

3.1% O_2
−2.2 inch H_2O

2.6% O_2
−1.6 inch H_2O

1.4% O_2
−1 inch H_2O

0.6% O_2
−0.5 inch H_2O

Air

F.D. fan

Fuel
gas

Figure 20.10 The "puffing" boiler: schematic drawing.

in-leakage in the convective section, it may sometimes experience thumping, puffing, or vibration.

Take a look at my sketch of a "puffing" boiler, Fig. 20.10. It is operating with perhaps a little too much draft, but look at the firebox oxygen; it is only 0.6 percent. However, the oxygen content of the flue gas steadily increases to 3.1 percent at the outlet from the convective section. This boiler has a tremendous amount of air leaking into the convective section. Perhaps someone is operating it on the basis of that 3.1 percent oxygen to the stack, without checking the firebox conditions—very foolish, as I hope you will now agree. This boiler is puffing, too; perhaps I should have shown its convective section walls alternatively convex and concave, but instead I have just entitled it "the puffing boiler." Here is what actually causes the puffing, vibrating, or thumping.

The fuel gas is not properly combusted at the burners, so partially combusted hydrocarbons and perhaps some totally unburned fuel pass up into the convective section with the flue gas. The convective section, as shown in Fig. 20.10, is under negative pressure, so air leaks in and mixes with the hot unburned hydrocarbons, which then reignite; thus we have afterburn in the convective section. The heat of the fire in the convective section expands in the flue gases, causing a positive pressure in the convective section, which stops the air leaks. After some period of time, the fire in the convective section has consumed all the tramp air and is extinguished. The convective-section flue gases then cool and shrink, reestablishing a negative pressure as before, and, as before, air leaks into the convective section and the hot, unburned hydrocarbons from the firebox mix with that tramp air and reignite in the convective section.

This cycle of events will continue, except each time the unburned hydrocarbons ignite in the convective section and the pressure changes from negative to positive the leaks will increase in size as the boiler fabric is shaken.

If the airflow to the firebox becomes very severely low, then one or more of the burners may be extinguished through lack of oxygen. If the fuel is then left to flow unchecked, it will build up in the firebox until eventually enough tramp air has been sucked into the firebox, at which point the extinguished burners may be reignited by the heat left in the refractory walls, causing a minor explosion or a detonation as it ignites. The magnitude of the explosion depends on the degree of oxygen starvation.

If anything like this is happening to your fired equipment, open up the air registers, get more air into the firebox, and patch the leaks!

Lighting burners

Before lighting burners in a piece of fired equipment that has been idle or off line for any time or whose burners have become extinguished during operation, it is vital to first purge the firebox of hydrocarbons and combustibles or explosive mixtures. Here is a method as outlined to me by a prominent burner manufacturing company:

(1a) If the furnace or boiler has a forced-draft fan, open the dampers and displace five volumes of air

(1b) If the furance or boiler is natural-draft, open the air register and stack damper and wait for 5 to 10 min.

2. Check the firebox with an hydrocarbon detector (the plant's fire station will usually have one of these and someone trained in its use). Pay particular attention to stagnant areas of the firebox, as fuel-gas valves tend to leak through burners and gases build up in areas of low airflow. Look for unburned gas and explosive mixtures. If you detect any such fuel or explosive mixtures, repeat step 1a or 1b, then test again with the hydrocarbon detector. Continue this process until the firebox is purged of all combustible or explosive material.

3. Light pilot lights.

4. Light burners using pilot lights.

Please note that it is not only bad practice but extremely dangerous to light main burners without the use of a properly installed pilot light. Butane torches, "fireworks," torches consisting of gasoline-soaked rags on sticks, or residual heat in the furnace refractory *do not* provide a safe alternative to a properly installed pilot light. Please also note that fired equipment either without burner pilot lights or with extinguished burner pilot lights has been known to explode during operation, as the pilots, when lit, do provide some means of safely relighting or keeping main burners alight. For this reason we prefer to install pilot lights rather than spark ignitors.

Today, burner manufacturers have access to all kinds of designs for pilot lights, to fit all shapes and sizes of burners and burner ports and draft systems from the smallest breeze to gale-force winds—there is no longer any excuse to operate without pilot lights installed and working.

Heat balancing

Let us finally look at a situation where one might wish to transfer heat duty from the firebox to the radiant section. At the beginning of the discussion on air preheaters earlier in this chapter, we noted how fitting an air preheater to an existing furnace would increase the flame temperature and possibly overheat the firebox.

Suppose a fired heater in an oil refinery, in service to heat crude oil, is fitted with a new air preheater. The firebox was running hot before the air preheater was installed, but the management was convinced that the installation of the air preheater would allow them to heat more crude oil. When the heater with its new air preheater came back on line, the following changes were noted:

- Fuel-gas consumption was down by 10 percent (for the same heat absorbed as before the revamp).

- The firebox, which was hot before the revamp, was now hotter, even at the original crude-oil flow rate; thus, rather than running more crude, operations had to cut back the crude rate to maintain satisfactory firebox conditions.

- Crude-oil side convection outlet temperatures were lower.

Here is what happened. Preheating the air by 300°F had raised the burner flame temperature by 300°F. The hotter flames then radiated more heat per pound of fuel consumed, and as a result, the firebox became much hotter. The same factor that reduces fuel consumption in the radiant section also reduced the flow of flue gas to the convective section, thus reducing the convective-section outlet temperature.

So what can we do to actually run more crude, especially as there is no bypass around this new air preheater?

The answer, of course, is to allow more air into the firebox, and thus generate more flue gas. The firebox oxygen was increased from 3 to 6 percent, which reduced the firebox temperature to its prerevamp state. This, in turn, increased the pounds of flue gas flowing through the convective section and increased the heat absorbed in the convective section.

The increased combustion airflow was needed not for combustion, but to transfer heat from the radiant section (firebox) to the convective section. This is what we call *heat balancing*. In this situation, oxygen requirements to reach absolute combustion become irrelevant as we are now operating with a very plentiful supply of oxygen.

21

Fired Heaters: Process Side

Coking Furnace Tubes and Tube Failures

The heat of combustion in a process heater or steam boiler may liberate 100×10^6 Btu/h. If the heater's efficiency is 78 percent, the heat absorbed by the tubes should be 78×10^6 Btu/h. When observing a heater's operation, it is a good idea to check this heat balance by calculation.

Process Duty versus Heat Liberation

The heat of combustion is a product of the amount of fuel consumed and the net heating value of the fuel. The heater's efficiency is a function of the flue-gas stack temperature, the excess air or oxygen, and the ambient-heat losses from the firebox and the convective-section structures.

The heat absorbed by the tubes is the sum of the heat of vaporization of the process liquid to vapor, plus the increase in the sensible-heat content of the flowing process fluid.

If the heat picked up from the combustion of the fuel does not equal the heat absorbed by the process fluid, then something is amiss with the data. Often, determining the cause of such an inconsistency will reveal several fundamental operating or measurement problems with a fired heater. Quite commonly, we may find that the metered fuel-gas rate is wrong, or that the ambient-heat losses are much greater than anticipated in our calculations.

Distribution of heat of combustion

When fuel is burned in a fired heater, there are three major products of combustion:

- Water
- Carbon dioxide
- Heat or British thermal units

The water, plus the CO_2, mix with the inert nitrogen in the combustion air to form flue gas. There is also some oxygen in this flue gas. This oxygen is called *excess* O_2. A typical excess O_2 content of flue gas is 2 to 6 percent. About 80 percent of the flue gas is N_2. The rest is H_2O plus CO_2.

The heat liberated by combustion is distributed in one of three ways:

- Some of the heat is used to increase the temperature of the combustion air from the ambient-air temperature, to the temperature of the flue gas leaving the heater's firebox. This heat is called the *convective-heat* content of the flue gas.
- Some of the heat is radiated directly to the heater tubes.
- The rest of the heat is radiated to the refractory walls.

In most heaters, the majority of the heat of combustion is radiated to the refractory walls. The glowing refractory walls then reradiate the heat to the heater tubes.

Many older heaters have massive brick refractory walls. The weight of these walls greatly exceeds the weight of the heater's tubes. The massive brick refractory walls store a great deal of heat. This creates several operating problems. One such problem is that it takes many hours to bring such a heater up to its normal operating temperature. However, a far more serious problem occurs when the process flow to a heater is interrupted.

When flow is first lost, the fuel to the heater is often automatically tripped off. But the refractory walls continue to radiate heat to the process tubes. It is rather like a large fly wheel. Even after we stop cranking the wheel, the rotational energy stored by the wheel keeps it spinning. Even when all the firing is stopped, the energy stored in the refractory walls continues to radiate heat to the process tubes.

But without flow, there is no way to carry the heat away from the tubes. Therefore, the tubes overheat. The temperature of the tubes may approach the temperature of the refractory, at the point in time when flow is lost. The refractory temperature is indicated by the firebox temperature, or the temperature of the flue gas flowing from the firebox into the convective section.

A typical firebox temperature is 1500°F. Thus, the heater tubes can reach 1300°F on loss of the process flow, even though the fuel flow has been immediately stopped. Tubes with a low chrome content may bend and distort as a result of such overheating. Even at 1000°F, residual liquid left in the tubes when flow is lost may thermally degrade to a car-bonaceous solid or heavy polymer that fouls the interior of the tubes.

One way to combat this problem is with steam. As soon as the flow is interrupted, high-pressure purge steam is automatically opened into the heater tube inlets. The steam blows the residual liquid out of the tubes, and also helps remove heat from the tubes.

Stuttering-feed interruption.   After the purge steam begins to flow, the refractory walls and the tubes slowly cool. However, if process flow is reintroduced to the heater during this cooling period, a serious problem may result. The first few gallons of process liquid flowing through the heater will become extremely hot. The liquid may get so hot that it will turn to solid coke, and partially plug the heater tubes.

If this problem—the sudden loss of flow, followed by the premature restoration of flow—occurs repeatedly over a period of a few hours, then layers of fouling deposits or coke are accumulated inside the tubes until a heater shutdown becomes unavoidable. This sort of failure is called a *stuttering-feed interruption.*

Modern technology has come a long way in mitigating such problems. In new heaters, lightweight ceramic tiles, rather than massive brick refractory walls, are the norm. These ceramic tiles do not store very much heat. Hence, when the heater process flow is reduced or lost, as long as the fuel flow is quickly curtailed, the tubes tend not to overheat.

Adiabatic combustion

In 1980, there was a strike at the Amoco Oil Refinery in Texas City. I was assigned during this emergency to work as the chief operator on the sulfur plant. You see, I was the engineer who had designed the sulfur plant. Therefore, Larry Durland, the refinery manager, thought that I would be the logical person for the chief operator's job. To be hon-est, I was a scab. I know I was a scab, because when the hourly opera-tors returned to work, they gave me a gift: a tee-shirt. On the front it said "SCAB." One the back it said "I like cheese," meaning I was a rat!

A sulfur plant converts H_2S to elemental sulfur, through the partial oxidation of H_2S:

$$H_2S + O_2 + N_2 \longrightarrow H_2O + N_2 + S \text{ (vapor)} \qquad (21.1)$$

This reaction takes place in an *adiabatic-combustion chamber,* shown in Fig. 21.1. This chamber has no tubes to absorb radiant heat.

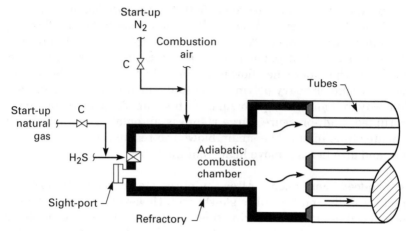

Figure 21.1   Front end of a sulfur recovery plant.

Plenty of radiant heat is liberated, but only to the refractory brick walls. The bricks then reradiate the heat back into the gaseous products of reaction. This is called *adiabatic combustion* because no heat is lost from the combustion reaction to radiation. The adiabatic-combustion temperature for the preceding reaction [Eq. (21.1)] is about 2300°F. The refractory used to contain this high temperature is manufactured from 90 percent alumina. Such refractory may be exposed to temperatures of up to 2900°F, without damage.

During the strike, the sulfur plant was shut down for minor repairs. I had to supervise its start-up. Mainly, I had to reheat the adiabatic-combustion chamber to 1800°F, before restoring the flow of H_2S. This was done by burning a controlled amount of methane or natural gas, with a carefully regulated flow of air. The idea was to slowly heat up the combustion chamber with hot flue gas by 100 to 200°F per hour. This slow reheat was needed to avoid cracking the refractory bricks, because of uneven heating. To carry away a portion of the heat of combustion of the natural gas, we used pipeline nitrogen.

I thought it best to control the flows of nitrogen, natural gas, and air, myself. Basically, the other scabs on the sulfur plant were head office people, whom I did not trust. The reheat phase of the start-up seemed to be going quite slowly. The combustion chamber temperature crept up by 50°F an hour, rather than the normal 150°F per hour. To speed the reheat, I reduced the nitrogen flow. This helped, but not by much.

It all seemed so odd. Especially as the interior of the chamber, viewed through the sight-port (see Fig. 21.1), had a dazzling white appearance. I remember thinking, "I cannot spend the rest of my life out here. I will have to speed this along." So I shut off the nitrogen, and turned up the gas and air. Finally the chamber temperature started to climb at a respectable rate. But then I began to see something in the chamber which was utterly impossible.

Of course, peering through the glass sight-port, it is hard to see very well. But it seemed to me as if the opposite refractory wall was very slowly, but still perceptibly—melting! But this could not be, for several good reasons:

- *Reason number one:* The 90 percent alumina refractory was rated for 3000°F service.

- *Reason number two:* The indicated combustion chamber temperature was still only 1500°F.

- *Reason number three:* I am a nice person, and bad things should not happen to nice people.

After 2 hours, it became clear that the openings in the opposite refractory wall were shrinking. These openings permitted the hot flue gas to exit the combustion chamber and flow into the tubes of a heat-recovery boiler. The apparently melting refractory was sagging and restricting these apertures.

I noticed, with sinking heart, that the combustion chamber pressure was steadily rising. It rose from 2 to 12 psig, at which point the fuel gas tripped off as a result of the high pressure. The sulfur plant start-up was aborted! But what had happened?

An investigation showed that the *thermowell*—a steel or ceramic tube, containing the thermocouple wires—was not fully inserted into the adiabatic-combustion chamber. The end of the thermowell was only half way into the 12-in-thick refractory wall, as shown in Fig. 21.2. Therefore, the thermocouple was measuring the relatively cool zone inside the refractory wall, rather than the far hotter zone in the combustion chamber. I should have paid closer attention to the physical appearance of the chamber. Bricks glowing bright red are radiating heat of about 1500°F. Bricks glowing a dazzling white are radiating heat above 2800°F. (see Table 21.1).

How hot did the refractory surface get? This value can be calculated if we assume the following:

- Fuel was 100 percent methane.
- Complete combustion (implying theoretically perfect air/fuel gas mixing).
- No excess oxygen.
- No heat losses to surroundings.
- Air is available at 60°F.

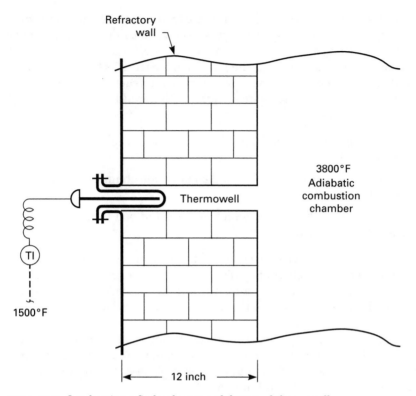

Figure 21.2 Overheating a firebox because of shortened thermowell.

TABLE 21.1   Visual Estimation of Temperatures

Color	Temperature, °F
Black red	990
Dark red	1050
Dark cherry	1250
Bright cherry	1375
Light red	1550
Orange	1650
Light orange	1725
Yellow	1825
Light yellow	1975
White	2200
Dazzling white	2730

Then, the calculated temperature of the combustion chamber is 3800°F. This is called the *adiabatic flame temperature*. Such a temperature is quite sufficient to turn even bricks into a high-viscosity, lava-type, semisolid fluid.

Larry Durland, the refinery manager, was not particularly pleased. He unjustly blamed me for this incident. Just because I had designed the sulfur plant, was the chief operator, and had been personally restreaming the sulfur plant, he said it was all my fault, and that I should have known better.

Heater Tube Failures

Heater tubes are designed to operate at a particular pressure and temperature. The design pressure of the tube is not the inlet operating pressure of the heater. The design tube pressure is the heater charge pump dead-head, or shut-in, pressure, as discussed in Chap. 23. The design temperature of the tube is not the heater outlet process operating temperature. The design tube temperature is the anticipated or calculated maximum *tube skin temperature* (at end-of-run conditions), which is simply the temperature of the exterior metal surface of the tube. Many plants now call this temperature the *tube metal indication* (TMI).

The calculated tube skin temperature is mainly a function of the fouling resistance assumed inside the tube. The greater the assumed fouling resistance, the higher the design tube skin temperature, and the thicker the tube wall. In a sense, then, we partially assume the design tube thickness, on the basis of experience, for a particular plant service.

A typical process heater tube diameter is 4 to 10 in. Tube thickness is usually between $\frac{1}{4}$ and $\frac{1}{2}$ in. Heater tubes are often constructed out of chrome steel. A high chrome content is 13 percent. The chrome content increases the heat resistance of the tube. A tube with a 11 to 13 percent chrome content can normally withstand a skin temperature of up to 1300 to 1350°F. A low-chrome-content tube of perhaps 3 percent may be limited to 1200°F tube metal temperature. Naturally, the pressure, thickness, and diameter of the tube all affect its maximum skin temperature limitations.

For added corrosion and temperature resistance, the nickel content of tubes, and sometimes the Moly (molybdenum) content as well, are increased. Tubes with a high nickel content are classified as 300 series stainless steels. A 0.5 percent silicon content is used to enhance the tube's oxidation or exterior scaling resistance.

High-temperature creep

When the tube metal temperature exceeds a value of 1300 to 1400°F, it becomes plastic. This means that the pressure inside the tube causes the tube diameter to expand. This is called *high-temperature creep.* As the diameter of the tube bulges and expands, the tube walls become progressively thinner and ultimately too thin to constrain the pressure inside the tube, and the tube bursts. Large-diameter tubes, operating at higher pressures and with a thin wall thickness, fail at a relatively low tube skin temperature.

Tubes seldom fail because of external oxidation, and tubes rarely "burn up." They fail because of high-temperature creep, which causes the tube to expand and burst. Thus, the fundamental cause of tube failure is a high localized temperature, which is called a "hot spot."

Purge steam

When a heater tube fails, the process fluid spills out into the firebox. Let's assume that the process fluid is a combustible liquid. Will the leaking process fluid burn? The answer is, mostly not. There is probably not enough excess oxygen in the firebox to support a substantial amount of additional combustion.

What will happen is that flames and black smoke will pour out of the heater's stack. It looks very bad, very dangerous, but it is really not. There is not enough excess oxygen in the firebox to cause a very high temperature. The combustion gases, or flue gas, are too fuel-rich to explode. We say that the flue gas is above the *explosive region.*

To prevent the flue gas from exploding, we need to proceed rather cautiously. If we just block in the leaking process tubes and fuel-gas supply, the hydrocarbon content of the flue gas will gradually decrease. The air:fuel ratio in the flue gas will increase until the flue-gas composition enters the explosive region. If the firebox refractory walls are still hot enough to initiate combustion, the firebox will now explode.

This happened on a unit I was supervising. A score of operators were lying on the concrete with their polyester shirts burning. This incident happened 20 years ago, before the need to wear fire-retardant Nomex coveralls was recognized.

The correct way to prevent this sort of firebox explosion is to use firebox purge steam. A typical heater firebox might have a half dozen 3-in purge steam connections. The idea is to displace the air in the firebox with steam. Then, the flows of combustible process liquid and fuel gas may be safely stopped, without fear of entering the explosive region. This purge steam is different from the purge steam used in the heater's tubes.

Identifying thin tubes and hot spots

A bulging tube indicates a thin area of a tube. If the diameter of a tube were to uniformly increase by 20 percent, then the thickness of the tube would decrease by 20 percent. But tubes rarely expand uniformly. They expand mainly on that side of the tube that is hottest. Hence, a tube bulge that increases the overall tube circumference by 20 percent typically reduces the thickness of the tube in the area of the bulge by 40 percent. For many tubes, this reduction in thickness is sufficient to cause tube failure. There is no theoretical basis for these statements. It is just a summary of what I have seen, when a failed section of tubing is cut out from a heater for failure analysis (See Fig. 21.3).

The best way to find thin, bulging tubes when the heater is off line is to run a ring down the length of a tube. A really severe bulge may be visually observed, even when the heater is in service. Sagging tubes do not represent thin tubes. I have seen radiant-section, horizontal tubes sag by more than one tube diameter and operate for years without failing.

Identification of hot spots.   Localized overheating of a tube causes localized high-temperature creep. This leads to the plastic deformation of a tube, and hence thin tube walls. Such hot spots are indicated by the color of the tube, as shown in Table 21.1. This chart is not a function of

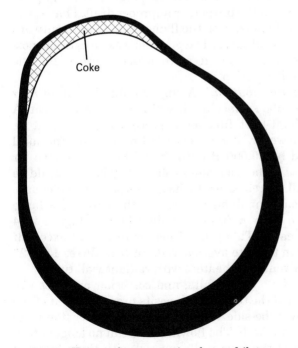

Figure 21.3   Heater tube cross section close to failure.

material of the tube. Carbon steel tubes, as well as tubes of all alloys, glow with a color which corresponds only to the localized tube temperature. This is the principle of operation of the *optical pyrometer,* a device that is widely used in the process industry to remotely measure tube skin temperatures.

Sometimes, deposits accumulate on the exterior of tubes. These deposits are unavoidable when heavy industrial fuel oil is fired.Such fuel oils contain high concentrations of vanadium, nickel, sodium, and iron (in that order of concentration). These metals deposit as an ash, on the exterior of both the radiant- and convective-section tubes. The ash will get very hot and will turn sections of tubes yellow, or even whitish silver. This is not an indication of tube overheating. If the deposits get thick enough, they will interfere with the rate of heat absorption by the tube. It is very difficult to distinguish between a real hot spot and glowing ash. Dirty fuel gas also leads to ash formation on the radiant surface of tubes.

What causes hot spots?   Three conditions will promote the formation of hot spots on tubes:

1. *Flame impingement.*   This is often caused by dirty burner tips, lack of combustion air, poorly designed burners, high burner tip pressure, improper adjustment of the burner, or improper draft. I have seen a heater in Cartagena, Colombia, with the flames being forced outward against the upper radiant wall tubes. The problem was an extreme positive pressure in the firebox, due to excessive pressure drop of the flue gas in the fouled convective section.

2. *Poor radiant-heat distribution.*   A high localized *radiant-heat flux* [in Btu/[(h)(ft^2)] of tube surface area] will result in a hot spot. A moderate average radiant-heat flux for a process heater might be 12,000 Btu/[(h)(ft^2)]. However, if heat distribution is poor, the local radiant-heat flux could be 20,000 Btu/[(h)(ft^2)]. This is excessive for most services. The cause of the poor radiant-heat distribution could be sagging tubes, too small a firebox for the heat released, or inadequate tube spacing. Having many small burners rather than just a few large burners promotes more even radiant-heat distribution. Figure 21.4 shows a double-fired heater. This type of heater results in excellent heat distribution and can tolerate average radiant-heat fluxes 15 to 25 percent higher than conventional heaters with radiant wall tubes.

3. *Interior tube deposits.*   Coke, salts, and corrosion products can adhere to the inside wall of the tube. The deposits precipitate out of the flowing fluid, due to low tube-side velocity. Once the solids stop moving, they bake onto the tube wall. The flowing fluid can no longer effectively cool that portion of the tube wall, covered with these internal deposits. The minimum acceptable tube-side velocity to retard solids

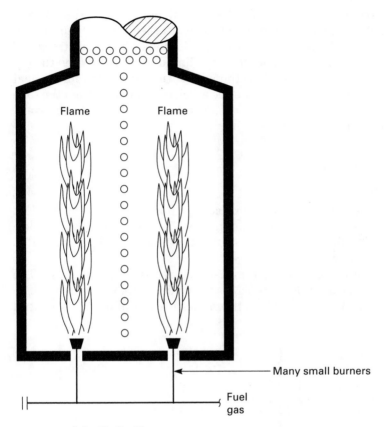

Figure 21.4 A double-fired heater.

sticking to the tube wall really depends on the service. A few general guidelines are

- Mixed-phase vapor and liquid flow, 20 ft/s
- Liquid-only flow, 8 ft/s.

When a tube that has failed because of a hot spot caused by a coke deposit inside is cut in half, we would observe a pattern similar to that shown in Fig. 21.3.

4. *Dry-point deposits.*   We sometimes see that a certain heater tube will glow a light red and fail for no apparent reason. The particular tube position in the firebox seems to be far more subject to failure than its neighbors. The tube is not located in an area of flame impingement, and the tubes upstream and downstream are a nice dark red (see Table 21.1).

This strange behavior is likely due to the heater feed going through its dry point in this tube position. Let's say that the heater feed is 10 weight percent (wt%) liquid. The heater effluent is 100 percent vapor. Unfortunately, the liquid portion of the heater feed is likely to contain small amounts of dissolved, or suspended, solids. The concentration of these solids increase in the liquid as the liquid vaporizes. When the liquid dries out, the solids stick to the wall of the tube. The radiant heat absorbed by the tube can no longer be efficiently removed from the tube wall by the flowing vapor. The tube overheats and bulges, the walls thin at the bulges, and the tube bursts.

Flow in Heater Tubes

Loss of flow

Let us say we have an ordinary orifice-type flowmeter, as shown in Fig. 21.5 (see Chap. 6, "How Instruments Work"). What happens if the low pressure (i.e., the downstream) orifice tap plugs? Does the indicated flow go up or down?

If a pressure tap plugs, the measured pressure will decrease. The measured pressure difference across the orifice plate will increase. The indicated flow will go up. If the flowmeter is controlling a flow-control

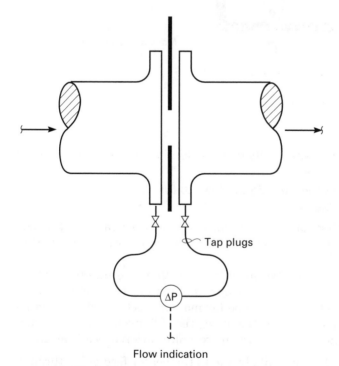

Flow indication

Figure 21.5 Plugged low-pressure tap increases indicated flow.

valve on a tube inlet to a heater, the valve will then close. Flow to the tube inlet will be lost. The tubes downstream of this flow-control valve will likely be damaged because of overheating, or they will plug and foul, as a result of thermal degradation of the process fluid.

One way to limit the damage due to this all-too-common problem is to limit the amount the flow-control valve can close on automatic control to, say, 20 percent open. Make sure, though, that with the valve 20 percent open, enough flow is sustained through the tubes to prevent tube damage due to too low a process flow.

Annular flow

Sometimes, the liquid flow through a heater tube is low. Low flow promotes thermal degradation of the liquid and harmful buildup of deposits inside the tubes. To combat this problem, we often add *velocity steam* to the heater's tubes, to increase the linear velocity of the oil. However, if the physical properties, especially the density, of the oil and steam are very different, phase separation will result. This means that the low-density steam will run down the center of the tube. The higher-density liquid will creep along the periphery of the tube, at a much lower velocity. This sort of phase separation is called *annular flow*. When a tube is in this flow regime, increasing steam velocity further does very little to increase the liquid velocity, or in combating the thermal degradation of the liquid.

Low-NOx Burners

NOx stands for a variety of nitric oxides. Many heaters in the United States have been retrofitted with staged burners in the last decade or so. These staged burners combust the fuel in two or three stages. For example, in the burner shown in Fig. 21.6, 50 percent of the fuel is burned with 100 percent of the air. The flame produced by this first stage of combustion radiates heat to the process tubes and refractory walls. Next, the remaining 50 percent of the fuel is added around the circumference of the first-stage burner. This second stage of combustion also liberates radiant heat. But because the radiant heat is liberated in two steps, the maximum flame temperature is reduced. This has two favorable results:

1. Nitric oxide (NOx) production is reduced, as the oxidation of nitrogen is a strong function of the flame temperature. Hence the term low-*NOx burners*.

2. Heat liberation from the low-NOx burner is more uniform than with a conventional burner. This permits a higher average firebox tem-

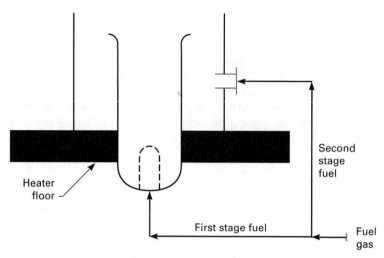

Figure 21.6   A two-stage low NOx burner.

perature to be sustained, without promoting hot spots on the tubes and interior tube coking. I have recently seen a refinery crude unit vacuum tower feed heater expanded by 15 percent, simply by retrofitting with two-stage, low-NOx, fuel-gas burners.

These burners replaced old-fashioned burners that did a poor job of mixing the air and fuel gas.

Premix, or primary-air, burners do a great job of mixing air and fuel gas. But they also produce a high flame temperature and hence, higher concentrations of NOx in the heater's effluent flue gas.

Fire-Side Heaters

So far, we have been discussing fired heaters, with the fire outside the tubes; that is, the fire is outside the heat-exchange surface. Many fired heaters have the fire, or at least the hot flue gas, inside the heat-exchange surface.

One common example of firing on the tube side of a heater is the gly-col-regeneration boiler, shown in Fig. 21.7. This type of heater will typically have a high excess O_2, to prevent high flame temperatures, which could overheat the fire tube. Also, the fire tube is kept submerged in liquid, to prevent tube overheating.

In multiple-tube boilers (usually horizontal) the fire may be on the tube side. As long as the fire tubes are kept submerged in water, the tubes do not overheat. Boilers of this type are widely used in the regen-

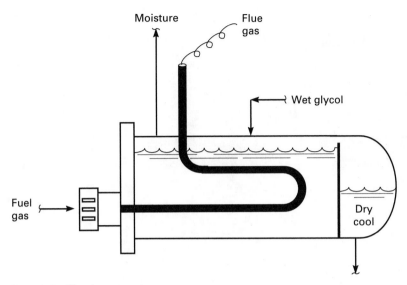

Figure 21.7 Glycol regenerator.

eration of spent sulfuric acid, and in the production of elemental sulfur from hydrogen sulfide.

The great difficulty with such fire-side, multitube boilers is over-heating the hot-side tubesheet. This tubesheet is exposed to the full temperature of the combustion gases, but is difficult to keep cool with the boiler's feedwater. Also, the ends of the tubes, where they are sealed in the tubesheet, are hard to keep cool. To protect these ends from the direct radiant heat in the combustion chamber, ferrule inserts, about $1\frac{1}{2}$ in long, are cemented into the front end of each tube.

If these ferrules fail, or if the tubesheet overheats, the ends of the tubes will pull away from the tubesheet. The result is called a "roll leak." Boiler feedwater will blow out of these roll leaks. Rerolling and seal welding the ends of the tubes, while difficult, is then the only way to stop such leaks.

Refrigeration Systems

An Introduction to Centrifugal Compressors

Waste heat must occasionally be removed from a process at below-ambient temperatures. This is normally accomplished by evaporating a light fluid called a *refrigerant.* Common industrial refrigerants are

- Ammonia
- Propane
- Freon

Your home central air-conditioning system is a good example of a simple refrigeration loop. Figure 22.1 illustrates the basic components:

- Refrigerant motor-driven compressor
- Refrigerant condenser
- Refrigerant letdown valve
- Evaporator
- Refrigerant receiver

The basic process flow of the refrigerant system is

1. The compressed refrigerant vapor is discharged from the compressor. This vapor is *superheated,* meaning that it is above its dew-point temperature.

2. The vapor is next cooled and condensed in the refrigerant condenser. Typically, 75 to 80 percent of the condenser heat duty is the

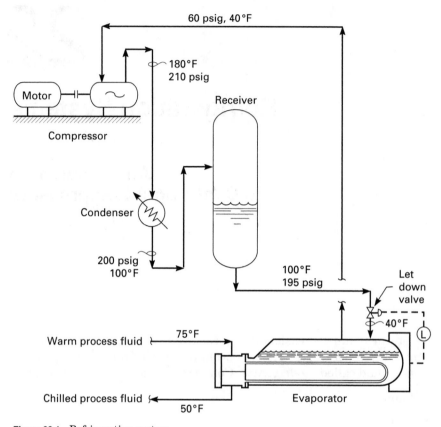

Figure 22.1 Refrigeration system.

latent heat of condensation of the refrigerant. The remainder is sensible-heat removal. The sensible heat removed is a combination of desuperheating the vapor and subcooling the condensed liquid refrigerant.

3. The subcooled refrigerant next flows into the refrigerant receiver. *Subcooled* means that the refrigerant temperature is below its boiling point, or bubble-point temperature. The liquid refrigerant loses a small amount of pressure as it enters the receiver. The pressure loss is due to friction in the piping and an increase in the elevation of the refrigerant as it enters the receiver.

4. The condensed liquid refrigerant next flows through the *letdown valve*. In Fig. 22.1, this letdown valve is not controlling the receiver liquid level. The receiver level is allowed to vary as the total refrigerant inventory changes. This idea can be better understood if we remember that this is a closed system. That is, we have to take swings in the refrigerant inventory somewhere in the system.

5. The refrigerant liquid partially flashes to a vapor as it flows through the letdown valve. The flashing represents the conversion of the sensible heat of the refrigerant to latent heat of vaporization. In Fig. 22.1, the refrigerant is chilled from 100 to 40°F. Approximately 25 percent of the liquid flashes to a vapor to provide this autorefrigeration.

The compressor, motor, condenser, receiver, and letdown valve are all components of your home central air-conditioning unit. They are installed as a package, surrounded by the condenser, outside your house. The evaporator is located in your attic. To continue our description of the process flow:

- The partially vaporized refrigerant flows into the evaporator. In Fig. 22.1, the evaporator shown is similar to a kettle-type reboiler (see Chap. 5). The process fluid flows through the tube side of the kettle evaporator. The refrigerant liquid level is maintained by the letdown valve. The refrigerant vapor flows from the top of the kettle, to the compressor suction.
- The motor-driven compressor boosts the pressure of the refrigerant vapors from the evaporator pressure, up to the condenser pressure.

Refrigerant Receiver

Looking at Fig. 22.1, the need for the refrigerant receiver is not immediately obvious. It does not appear to serve any process function. Many refrigerant systems are constructed without this vessel. However, it does have several important uses. For one thing, it provides a quiet zone for entrained lubricating and seal oils to settle out.

Lubricating and seal oil often leak into the circulating refrigerant. The lower the suction pressure of the compressor, the greater the problem. Eventually, the amount of heavy oil that accumulates in the evaporator will cause a problem. The heavy oil is largely insoluble in the refrigerant. As the refrigerant boils away out of the evaporator, the heavy oil is left behind. Apparently, this oil partly coats the refrigerant tubes. The heat-transfer efficiency of the evaporator is reduced.

Turbulence caused by the boiling of the refrigerant prevents efficient draining of heavy oil from the evaporator. However, if the refrigerant receiver has been designed properly, it can be used to trap out the heavy oil.

Inventory control

In Fig. 22.1, there are two levels that must be controlled. The evaporator level and the refrigerant receiver level. It is impossible to control both levels with two level-control valves. One level must be held with a level-control valve. The other level must be held by adjusting the system liquid refrigerant inventory. It is usually better to add makeup refrigerant to keep a decent liquid level in the receiver, and control the flow of refrigerant into the evaporator. However, reversing this control scheme by adjusting inventory control in the evaporator is also possible.

Vapor trap

Another way of looking at the receiver vessel is to think of it as a giant steam trap. As long as the liquid level is maintained in the receiver, uncondensed refrigerant vapors cannot escape from the condenser. Let's assume that such vapors did escape from the condenser. If there were no refrigerant receiver vessel, these vapors would blow through the letdown valve. Reducing the pressure of the vapors produces just a tiny bit of refrigeration as compared to vaporization of liquefied refrigerant.

Now let's assume that there is a refrigerant receiver between the condenser and letdown valve. This traps any uncondensed vapors. The accumulation of these vapors raises the pressure in the receiver. This puts backpressure on the condenser. The higher condenser pressure promotes more complete condensation of the refrigerant vapors.

Finally, the receiver will accumulate any noncondensable (or hard-to-condense) components that have accidentally entered the system. Air left in the vessels on start-up is one such example. Traces of methane and ethane in a propane refrigerant system is another. These light vapors may be vented from the top of the receiver during normal operations.

Evaporator Temperature Control

The system shown in Fig. 22.1 has no provision for controlling the evaporator temperature; this is controlled solely by the compressor suction pressure. The lower the compressor suction pressure, the lower the evaporator temperature. This is exactly how our home air conditioner works. If the evaporator temperature is too cold, what can we do? Well, if this were a steam turbine compressor, gas engine drive, or any other type of variable-speed driver, we could reduce the compressor's speed. This would reduce the flow of refrigerant, and raise both the evaporator and compressor suction pressure.

But, we are working with an ordinary AC (alternating-current) motor—which is a fixed-speed device. There are then three methods available to control the temperature in the evaporator:

Spillback

Spillback bypasses the compressor discharge gas, back to the compressor suction. This is a relatively energy-inefficient way to increase the evaporator temperature. Spillback is discussed in Chap. 28, "Centrifugal Compressors and Surge."

Discharge throttling

Discharge throttling is also an energy-inefficient way to control the evaporator temperature. Even worse, if we are working with a centrifugal compressor, discharge throttling may cause the compressor to *surge*.

Suction throttling

Suction throttling is the preferred method, as shown in Fig. 22.2.

Suction throttling also wastes energy, however; whenever a control valve is partially closed in any service, energy is always wasted. But with a fixed-speed centrifugal compressor, it is still the best option. As the suction valve is closed, two things happen. The flow of refrigerant vapor to the compressor is reduced, and the compressor suction pressure drops. The reduction in the refrigerant vapor flow reduces the workload on the compressor. But the reduction in the suction pressure, increases the *compression ratio* (discharge pressure divided by suction pressure). This increases the workload on the compressor.

As we will discuss in Chap. 28, increasing the compression ratio is a small effect on compressor workload, as compared to decreasing the vapor flow. Hence, suction throttling will significantly reduce the horsepower load on the compressor, as well as increase the evaporator temperature.

Suction throttling is usually done with a "butterfly" control valve, which has a very low pressure drop when it is wide open.

Compressor and Condenser Operation

Compressor operation

Let's say we have a centrifugal refrigeration compressor, driven by a motor. The motor is tripping off because of high amperage. Should we open or close the suction throttle valve shown in Fig. 22.2? Answer—close it. Of course, both the evaporator vapor outlet temperature and the process fluid outlet temperature will increase. But that is the price we pay for having too small a motor driver on the compressor. Does this mean that when our home air conditioner gets low on freon, our electric bill drops? Correct. But the price we pay is a hot home.

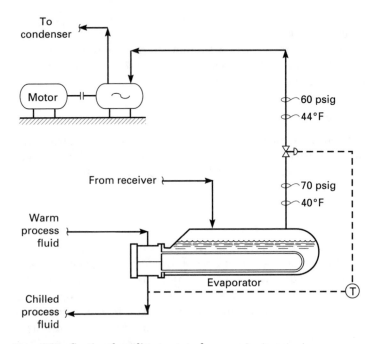

Figure 22.2 Suction throttling to control evaporator temperature.

How about the compressor discharge pressure? What is this controlled by? Answer—the condenser. The compressor discharge pressure has nothing whatsoever to do with the compressor. And this statement is true regardless of whether the compressor is a reciprocating or centrifugal machine, or fixed- or variable-speed. It is the condenser temperature that determines the compressor discharge pressure.

Lowering the compressor discharge pressure can be achieved by changing to cooler water flowing to the refrigerant condenser. But how does this affect the horsepower load or electrical power demand on the compressor's driver? Well, it depends. It depends on how much the refrigerant vapor flow increases, as the compressor discharge pressure is reduced.

For instance, suppose we are working with a reciprocating compressor. This particular machine has a very small volumetric clearance and a very high volumetric efficiency (these mysterious terms are fully explained in Chap. 29, "Reciprocating Compressors"). With this compressor, a significant decrease in discharge pressure will result in only a tiny increase in vapor flow. Therefore, the load on the driver will decrease.

On the other hand, suppose we are working with a centrifugal compressor. This particular machine has a very flat operating curve. This means that a small decrease in discharge pressure will result in a large increase in vapor flow. Therefore, the load on this compressor's driver will increase.

This discussion assumes that the compressor's suction pressure is fixed. However, in the real world, this is rarely the case. As we lower the compressor discharge pressure, the compressor suction pressure will also decrease. This will have the normally beneficial effect of reducing the evaporator pressure and temperature.

Now that the evaporator is cooler, we could increase the flow of warm process fluid to the evaporator. Thus, improving the efficiency of the condenser will increase the capacity of a refrigeration system to cool. This happens mainly by lowering the compressor discharge pressure rather than producing a colder condensed refrigerant.

Condenser operation

A closed-loop refrigerant condenser ought to be one of the cleanest services in a process plant. Even seal- or lube-oil leaks affect the evaporator efficiency, rather than the condenser. I have measured rather high [e.g., 140 Btu/[(h)(ft^2)($°$F)]] heat-transfer coefficients in such condensers, even after that condenser has been in service for several years since its last cleaning.

The most common problem with refrigeration condensers, is condensate backup, rather than fouling (see Chap. 13). This may happen if the condenser outlet line is undersized. Let's say that the pressure drop, due to piping friction losses in this line, is 10 psig. The refrigerant is isobutane. The isobutane cannot flash as it enters the refrigerant receiver. The reason it cannot partially vaporize is that there is no vapor outlet from the receiver. If it were to vaporize, it would slightly increase the receiver pressure. This would push up the liquid level in the condenser. More of the tubes in the condenser would be submerged in the liquid refrigerant. The refrigerant would be subcooled below its condensation or bubble-point temperature. It would have to be subcooled, so that after the liquid lost 10 psig of pressure, it would not flash.

To subcool isobutane to prevent it from vaporizing after losing 10 psig, the isobutane vapor pressure must be reduced by 10 psig. This would, in turn, require cooling the isobutane by about 5$°$F (see Fig. 9.1 in Chap. 9). Unfortunately, subcooling a liquid is difficult. This means that the heat-transfer coefficient for subcooling might be 30 Btu/[(h)(ft^2)($°$F)]. Even though the amount of heat to be removed in subcooling a refrigerant is negligible, it may require 30 percent of the tubes in the condenser to do the job. The reason is the low heat-transfer coefficient for subcooling, as compared to the high heat-transfer coefficient for condensation.

The tubes in the condenser required for subcooling steal heat-transfer surface area required for condensation. In effect, the condenser shrinks. This makes it more difficult to liquefy the refrigerant vapor. The vapor is then forced to condense at a higher temperature and pressure. Of course, this raises the compressor discharge pressure. And, as we have seen in a previous section, this increase in compressor discharge pressure invariably reduces the compressor's capacity and may also increase the horsepower needed to drive the compressor.

An increase in the elevation that the condenser effluent must flow up into the receiver will reduce the pressure of the liquid refrigerant, in the same way as the 10-psig piping friction losses. This loss in pressure, due to increased elevation, will also require an increase in the surface area of the condenser sacrificed for subcooling.

Refrigerant Composition

A very large percentage of refrigeration systems are not limited by either the compressor driver horsepower, the condenser, or the evaporator. Many refrigeration systems are limited by either:

- Compressor suction volume limits
- The mechanical pressure rating of the compressor case or the refrigerant condenser shell.

If the compressor is suction-volume-limited, we say that it is *speed-limited*. If the compressor is driven by an AC motor, its speed is determined by the motor speed and any connecting gears. If the compressor is driven by a turbine, then the rated speed of the turbine limits the compressor speed.

Overcoming speed limits

While the actual cubic feet of refrigerant vapor to the compressor may be limited by the compressor's speed, we can still overcome this bottleneck. The trick is to make the refrigerant lighter or more volatile. A lighter refrigerant will allow the evaporator, shown in Fig. 22.1, to operate at a higher pressure and/or a lower temperature. Increasing the suction pressure increases the density of the refrigerant vapor flowing to the compressor. Even though the volume of gas flow is constant, the weight flow of the circulating refrigerant can be increased. Yes, we can get 6 lb of manure in a 5-lb bag, if we make the manure denser.

For example, adding lighter propane to heavier isobutane in a refinery alkylation unit is a common practice to increase refrigerant circulation. But this will work only when the compressor driver has spare capacity.

Of course, in process operations, there are no free lunches. Spiking propane into isobutane also makes the lighter refrigerant more difficult to condense. This will raise the compressor discharge pressure. Also, the operating pressure on the shell side of the condenser will increase. If either piece of equipment does not have an adequate design pressure, then increasing the volatility of the refrigerant cannot be permitted.

Pressure rating limits

The simplest way to lower the compressor discharge pressure is to reduce the cooling-water temperature to the condenser. One neat trick I have used is to use the cooling-tower makeup water on a once-through basis, to the refrigerant condensers. This supply of water is almost always 10 to 20°F cooler than the cooling-water supply.

Another way to reduce the compressor discharge pressure is to render the refrigerant heavier, or less volatile. Adding isobutane to a propane refrigerant is one common example. Naturally, this will lower the suction pressure to the refrigerant compressor shown in Fig. 22.1, and increase the actual volume of vapor flow. If the compressor is speed-limited, this will not be practical.

Adjusting the refrigerant composition does not significantly affect the amount of horsepower needed to provide a fixed amount of cooling at a certain evaporator temperature. This statement is also true, even if the type of refrigerant used is changed completely. Propane, isobutane, freon, and ammonia all have roughly the same refrigeration efficiency.

23

Centrifugal Pumps: Fundamentals of Operation

Head, Flow, and Pressure

Most of the pumps we see, both at home and at work, are centrifugal pumps. And most of these pumps are driven by constant-speed, alternating-current motors.

Head

Centrifugal pumps are *dynamic machines,* which means that they convert velocity into feet of head.

To explain this concept of converting speed or velocity into feet of head, let's look at Fig. 23.1. This is the water intake section of the Chicago Water Treatment Plant. Water flows from Lake Michigan into a large concrete sump. The top of this sump is well above the level of Lake Michigan. The line feeding the sump extends three miles (3 mi) out into the lake. The long line is needed to draw water into the plant, away from the pollution along the Chicago lakefront. The pipeline diameter is 12 ft. Water flows in this line at a velocity of 8 ft/s. Six pumps, stationed atop the concrete sump, pump water into the Water Treatment Plant holding tanks.

One day, the plant experiences a partial power failure. Three of the six pumps shown in Fig. 23.1 shut down. A few moments later, the

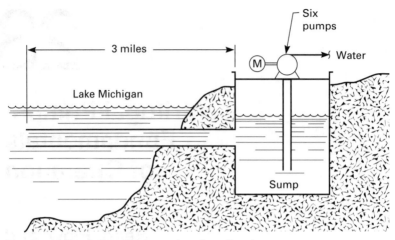

Figure 23.1 Converting momentum into feet of head.

manhole covers on top of the sump blow off. Geysers of water spurt out of the manholes. What has happened?

Hydraulic hammer

This is an example of water, or hydraulic, hammer. But what causes water hammer?

Let's consider the change to the velocity in the 3-mi-long pipeline, when half the pumps failed. The velocity in the pipeline dropped from 8 down to 4 ft/s. *Velocity* is a form of energy, called *kinetic energy*. Energy can take several different forms:

- Heat (Btus)
- Elevation or potential energy
- Kinetic energy (miles per hour)
- Pressure (psig)
- Electrical power (amperes)
- Work (horsepower)
- Acceleration
- Chemical (heat of reaction)

About 250 years ago, Daniel Bernoulli first noticed two important things about energy:

- Energy in one form can be converted to energy in another form.
- While energy can be changed from one form to another, it cannot be created or destroyed.

This idea of the conservation of energy is at the heart of any process plant. In the case of the Chicago Water Treatment Plant, the reduced velocity of the water was converted into feet of head. That is, the elevation of the water in the sump suddenly increased, and blew the manhole covers off the top of the sump.

Momentum

If the length of the pipeline had been a few hundred feet, this incident would not have happened. It is not only the sudden reduction of the velocity of the water that caused an increase of the water level in the sump. It is also the mass of water in the 3-mi-long pipeline that contributed to the increased height of water in the sump. The combined effect of mass times velocity is called *momentum.*

The mass of water in our pipeline weighed 160×10^6 lb. This much water, moving at 8 ft/s, represents a tremendous amount of energy (about 500 million Btu per hour). If the flow of water is cut in half, then the momentum of the water flowing in the pipeline is also cut in half. This energy cannot simply disappear. It has to go somewhere. The energy is converted to an increase of feet of head in the sump; that is, the water level in the sump jumps up, and blows the manhole covers off the top of the sump.

Incidentally, this is a true story. Can you imagine what would have happened if all six pumps failed simultaneously? The result would be a dramatic lesson in the meaning of water hammer.

My washing machine

Figure 23.2 is a picture of our washing machine as it was originally installed. Whenever the shutoff valve on the water supply line closed, water hammer, or hydraulic shock, would shake the water piping. The momentum of the water flowing in the piping would be suddenly converted to pressure. If the end of a piping system is open (as into a sump), then the momentum of the water is converted to feet of head. But if the end of the piping system is closed, then the momentum of the water is converted to pressure.

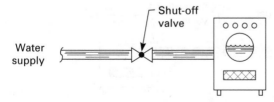

Figure 23.2 Hydraulic hammer hits home.

To fix this problem, I installed the riser tube shown in Fig. 23.3. The top of the riser tube is left full of air. Now, when the water flow in the supply pipe is shut, the momentum of the water is converted to compression energy. That is, the air in the riser tube is slightly compressed, as indicated by the pressure gauge I installed at the top of the riser tube.

Acceleration

Let us imagine that the six pumps in Fig. 23.1 have not run for a few days. The water level in the sump and the level of Lake Michigan will be the same. I now start all six pumps at the same time. An hour later, the water level in the sump is 12 ft below the water level in the lake. This 12 ft is called "feet of head loss." If the pipeline is 3 mi long, we say we have "lost 4 ft of head per mile of pipeline."

If we look down into the sump, what would we see happening to the water level during this hour? Figure 23.4 is a graph of what we would observe. The water level in the sump would drop gradually to 15 ft below the level in the lake. Then, the water level in the sump would come partly back up to its equilibrium level of 12 ft below the lake level. Why?

The loss of 12 ft of head as the water flows through the pipeline is due to friction; that is, 12 feet of head is converted to heat. But why do we have a temporary loss of an extra 3 ft? The answer lies in the concept of acceleration.

Let's say you are driving your car onto the expressway. To increase the speed of your car from 30 to 65 mph, you press down on the accelerator pedal. Having reached a velocity of 65 MPH, you ease off the accelerator pedal to maintain a constant speed. Why? Well, according to Newton's second law of motion, it takes more energy to accelerate your car than to keep it in motion.

Referring again to Fig. 23.1, the water in the 3-mi pipeline is initially stagnant. Its velocity is zero. An hour later, the water has accelerated to 8 ft/s. When you accelerate your car, the extra energy required comes

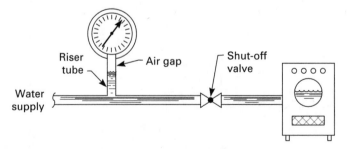

Figure 23.3 Riser tube stops hydraulic hammer.

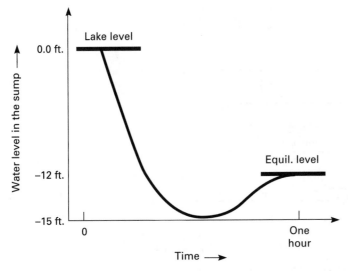

Figure 23.4 Effect of acceleration.

from the engine. But when we accelerate the water in the pipeline, where does the extra energy come from?

Does this extra energy come from the pumps? Absolutely not! The pumps are downstream of the pipeline and the sump. They cannot contribute any energy to an upstream pipeline. No, dear reader, the energy to accelerate the water in the pipeline must come from Lake Michigan.

But what is the only source of energy that the lake possesses? Answer—elevation or potential energy. To accelerate the water in the pipeline to 8 ft/s requires more energy than to keep the water flowing at that same velocity. And this extra energy comes from the 3 ft of elevation difference between the water level in the sump and the water level in the lake. Once the water has reached its steady-state velocity of 8 ft/s, the need for this extra conversion of feet of head to acceleration disappears, and the water level in the sump rises to within 12 ft of the lake's level.

Pressure

Loss of suction pressure

The need to accelerate liquid in the suction line of a pump leads to a difficult operating problem, which occurs on start-up. Just before the pump shown in Fig. 23.5 is put on line, the velocity in the suction line is zero. The energy to increase the velocity (i.e., accelerate) the liquid in the suction line must come from the pressure of the liquid at the pump's suction. As the pump's discharge valve is opened, the velocity

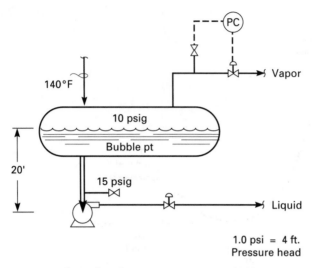

140°F

10 psig

Bubble pt

20'

15 psig

Vapor

Liquid

1.0 psi = 4 ft.
Pressure head

Figure 23.5 Loss of suction pressure causes cavitation.

in the suction line increases, reducing the pressure at the suction of the pump. The faster the discharge valve is opened, the greater the acceleration in the suction line, and the greater the loss in the pump's suction pressure.

If the pressure at the suction of the pump falls to its bubble or boiling point, the liquid will start to vaporize. This is called *cavitation*. A cavitating pump will have an erratically low discharge pressure and an erratically low flow. As shown in Fig. 23.5, the bubble-point pressure of the liquid, is the pressure in the vessel. We usually assume that the liquid in a drum is in equilibrium with the vapor. The vapor is then said to be at its dew point, while the liquid is said to be at its bubble point.

To avoid pump cavitation on start-up, the experienced operator opens the pump discharge valve slowly. Slowly opening the discharge valve results in reduced acceleration of the liquid in the suction line and a slower rate of the conversion of suction pressure to velocity.

Try this test. To illustrate what I have just explained, try this experiment:

1. Open the suction valve to a pump completely.

2. Crack open the case vent to fill the case with liquid. This is called "priming the pump."

3. Open the pump discharge valve completely.

4. Push the motor START button.

5. Observe the pump suction pressure.

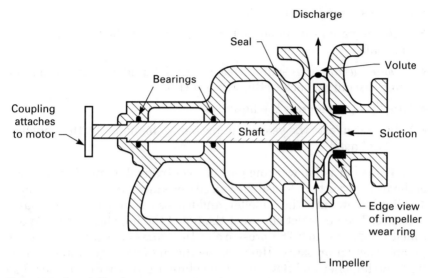

Figure 23.6 A centrifugal pump.

You will see that the pump suction pressure will drop, and then come partly back up. If the pump suction pressure were 15 psig to start with, it might drop to 12 psig and then come up to 14 psig. The permanent difference in the suction pressure between 15 and 14 psig is due to a 1-psig piping friction pressure loss. The temporary difference in the suction pressure between 14 and 12 psig is due to a 2-psig conversion of pressure to velocity or kinetic energy.

To review—all the energy needed to accelerate the liquid to the suction of a pump comes from the pump's suction pressure. None of this energy comes from the pump itself. Or, as one clever operator at the Unocal Refinery in San Francisco explained to me, "Pumps push, but they do not suck."

Pump discharge pressure

Figure 23.6 illustrates the internal components of an "overhung, single-stage" centrifugal pump. The term "overhung" refers to the feature that the pump has only an inboard, but no outboard, bearing. The inboard side of a pump, means the end closest to the driver. The term "single-stage" means that there is only one impeller. Multistage pumps can have five or six impellers.

The main components of the pump shown in Fig. 23.6 are

- *Shaft*—used to spin the impeller
- *Coupling*—attaches the shaft to the turbine or motor driver

- *Bearings*—support the shaft
- *Seal*—prevents the liquid inside the pump from leaking out around the shaft
- *Impeller wear ring*—minimizes internal liquid leakage, from the pump discharge, back to the pump suction
- *Impeller*—accelerates the liquid
- *Volute*—converts the velocity imparted to the liquid by the impeller, to feet of head

The impeller is the working part of a centrifugal pump. The function of the impeller is to increase the velocity or kinetic energy of the liquid. The liquid flows into the impeller, and leaves the impeller, at the same pressure. The black dot shown at the top of the impeller in Fig. 23.6 is called the "vane tip." The pressure at the vane tip is the same as the pump's suction pressure. However, as the high-velocity liquid escapes from the impeller and flows into the volute, its velocity decreases. The volute (which is also called the *diffuser*) is shaped like a cone. It widens out in the manner illustrated in Fig. 23.7. As the liquid flows into the wider section of the volute, its velocity is reduced, and the lost velocity is converted—well, not into pressure, but into feet of head.

Volumetric Flow

Feet of head

A centrifugal pump develops the same feet of head, regardless of the density of the liquid pumped, as long as the flow is constant. This statement is valid as long as the viscosity of the liquid is below 40 cp or 200 SSU (Saybolt Seconds Universal). But, as process operators or engineers, we are not interested in feet of head. We are interested only in pressure. Differential pressure is related to differential feet of head as follows:

$$\Delta P = \frac{(SG)\,\Delta H}{2.31}$$

where ΔP = increase in pressure, psig
ΔH = increase in feet of head, ft
SG = specific gravity of the liquid (i.e., the density of the liquid, relative to cold water)

The increase in pressure ΔP is also called *head pressure*. For example, if I have 231 ft of water in the glass cylinder shown in Fig. 23.8, it would exert a head pressure of 100 psig on the pressure gauge at the

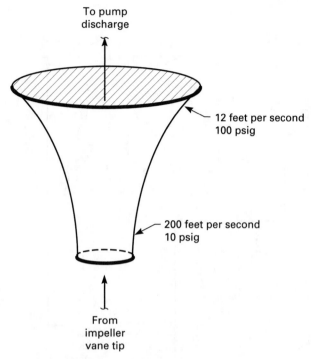

To pump
discharge

12 feet per second
100 psig

200 feet per second
10 psig

From
impeller
vane tip

Figure 23.7 A volute or diffuser converts velocity into feet of head.

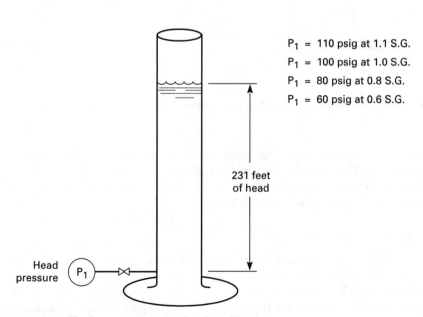

P_1 = 110 psig at 1.1 S.G.

P_1 = 100 psig at 1.0 S.G.

P_1 = 80 psig at 0.8 S.G.

P_1 = 60 psig at 0.6 S.G.

231 feet
of head

Head
pressure P_1

Figure 23.8 Head pressure compared to feet of head.

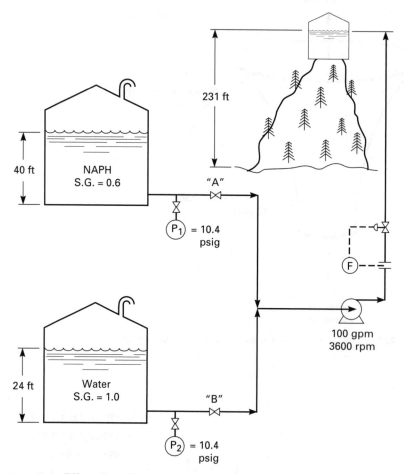

Figure 23.9 Effect of specific gravity on pump performance.

bottom of the cylinder. The head pressure would be proportionally reduced as the liquid becomes less dense. For example, 231 ft of kerosene (which has 0.75 SG) would exert a head pressure of only 75 psig, because it is lighter than water.

Effect of specific gravity

Figure 23.9 shows a centrifugal pump that may pump water through valve B or naphtha through valve A.

Question: When water is pumped, the pump's discharge pressure is 110.4 psig. The pump's suction pressure is 10.4 psig. When naphtha is pumped, what is the pump's discharge pressure?

Answer:

$$\Delta P = (110.4 - 10.4) \times \frac{0.60}{1.00} = 60 \text{ psig}$$

and the pump discharge pressure = 10.4 + 60 = 70.4 psig

Next question: The pump is driven by an ordinary alternating-current electric motor. It is running at 3600 rpm. When we switch from pumping water to naphtha, what happens to the speed of the pump?

Answer: Nothing. An AC motor is a fixed-speed machine. It will continue to spin at 3600 rpm.

Next question: The pump is pulling 100 amps of electrical power. The flow rate is 100 GPM. If we now close valve B and open valve A, what happens to the amp load on the motor driver?

Answer: The power demand will go down to 60 Amperes. Amp(ere)s are a form of electrical work. The units of work are foot-pounds. The feet of head developed by the pump is not affected by the specific gravity of the liquid. But the weight of liquid pumped is proportional to the specific gravity. If the specific gravity drops by 40 percent, and the liquid volume (GPM) stays constant, then the pounds lifted by the pump, drops by 40 percent and so does the electrical work.

Last question: Water is being pumped to a tank on a hill at 100 GPM. If we now switch to pumping a lighter fluid at the same rate, can we pump the lighter fluid to a higher elevation, only to a lower elevation, or the same elevation?

Answer: The same elevation. Centrifugal pumps develop the same feet of head at a given volumetric flow rate, regardless of the specific gravity of the liquid pumped. This means the ability of the pump to push liquid uphill is the same, even if the density of the liquid changes.

Pump curve

The feet of head developed by a pump is affected by the volume of liquid pumped. Figure 23.10 is a typical pump curve. As the flow of liquid from a centrifugal pump increases, the feet of head developed by the pump goes down, as does the pump discharge pressure.

A pump curve has two general areas. These are the flat portion on the curve and the steep portion of the curve. We normally design and operate a pump to run toward the end of the flat portion of its curve. A centrifugal pump, operating on the flat portion of its curve, loses only a small portion of its discharge pressure, when flow is increased. This is a desirable operating characteristic.

A centrifugal pump, operating on the steep portion of its curve, loses a large portion of its discharge pressure, when flow is increased. Pumps operating quite far out on their curve will have an erratic discharge pressure, as the flow through the pump varies. For one such pump, I

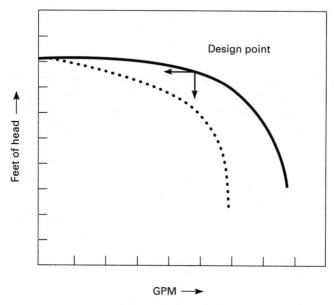

Figure 23.10 Centrifugal pump curve.

have seen the discharge pressure drop from 210 to 90 psig, as the flow increased by 40 percent. The flow through this pump was regulated by a downstream level-control valve. The discharge pressure varied so erratically that the operators thought the pump was cavitating.

Deviation from pump curve. Pumps always perform on their performance curve. It is just that they do not always perform on the manufacturer's performance curve. The dotted line in Fig. 23.7 compares an actual performance curve to the manufacturer's performance curve. Why the deviation?

- The manufacturer's curve was generated using water. We are pumping some other fluid, perhaps with a much higher viscosity.
- The clearance between the impeller and the impeller wear ring (Fig. 23.6) may have increased as a result of wear.
- The impeller itself may be worn, or the vane tips at the edge of the impeller might be improperly machined.
- If the pump is driven by a steam turbine, the pump speed could be lower than in the design.

Driver horsepower

Normally, increasing the flow from a centrifugal pump, increases the amperage load on the motor driver, as shown in Fig. 23.11. Driver

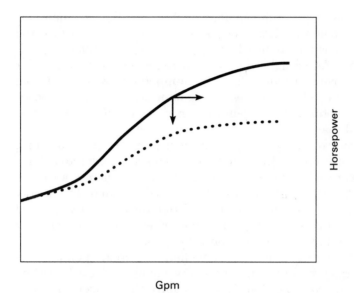

Gpm

Figure 23.11 Horsepower requirement for a centrifugal pump.

horsepower is proportional to GPM times feet of head. As shown in Fig. 23.10, as the flow increases, the feet of head developed by the pump decreases. On the flat part of the pump curve, the flow increases rapidly, while the head slips down slowly. Hence the product of GPM times feet of head increase.

On the steep part of the pump curve, the flow increases slowly, while the head drops off rapidly. Hence, the product of GPM times feet of head remains the same, or goes down.

I use the response of the amp load on the motor driver as an indication of pump performance. A pump operating on the proper, flat portion of its curve will pull more motor amps, as flow increases. A pump operating on the poor, steep proportion of its curve will not pull more amps. If the manufacturer's horsepower/flow curve indicates that more horsepower is required as flow is increased, but the motor amp load is not increasing, then something is wrong with the pump.

The amp meter for a pump is best located right on the start-stop box, next to the motor. Unfortunately, more often, the only amp meter is located on the electrical breaker box, which is at a remote location.

Pump Impeller

Impeller diameter

The pump impeller resembles a hollow wheel, with radial vanes. The diameter of this wheel can be trimmed down on a lathe. If an impeller with a 10-in diameter is trimmed down to 9 in, it will have a new performance curve, below its former performance curve. Both the flow and head of the pump will be reduced. Also, the amp load on the motor driver will decrease. As a rough rule of thumb, reducing a pump's impeller diameter by 10 percent will reduce the amp load by 25 percent. Not only will this save electricity, but the motor will run much cooler. Unloading the motor by 25 percent may increase the life of the motor windings by 10 years.

When the control valve downstream of a pump is operating in a mostly closed position, the upstream pump is a good candidate to have its impeller trimmed. Sometimes, the pressure drop across a control valve is so huge ($\geq$100 psi) that it makes a roaring sound. The energy represented by this wasteful ΔP is coming from the electricity supplied to the pump's motor. Hence, trimming the impeller also reduces wear because of erosion in downstream control valves.

We frequently increase the size of impellers to increase pump capacity and discharge pressure. However, many chemical plants break the law when they do so. State *boiler codes* require that vessels and heat exchangers, downstream of pumps, be rated for the

- Maximum pump discharge pressure
- At the maximum specific gravity that can reasonably be expected
- Unless the equipment is protected by a relief valve

Increasing the size of an impeller always increases its maximum discharge pressure. Hence, increasing the impeller diameter may be unlawful, depending on the *flange rating* of downstream heat exchangers.

Increasing the size of the impeller by 10 percent will increase the amperage load on the motor driver by 30 percent. For many process pumps, this would require a new motor and breaker, to support the larger impeller.

Trimming the pump impeller on a lathe also requires the reshaping of the tips of the impeller vanes. Most often, this requirement is ignored, and the revamped pump operates below its expected curve. Rather than worry about these fine points, it is best to order new impellers directly from the manufacturer.

Impeller rotation

Motor-driven pumps often are found running backward. If the motor is wired up backward, the pump will run with the impeller spinning backward. This will always reduce the pump's discharge pressure, sometimes by a little (10 percent) or sometimes by a lot (90 percent). It depends on the design of the impeller. You cannot see the direction of rotation of a pump, but if you touch a pencil to the spinning shaft, you can feel the direction of rotation. The correct direction of rotation, is indicated by an arrow stamped on the top of the pump case.

Effect of Temperature on Pump Capacity

We wish to increase the capacity of a centrifugal pump. Should we make the liquid hotter or colder? Let's make a few assumptions:

- Cooling the liquid will not increase its viscosity above 30 or 40 centistokes.

- We are pumping hydrocarbons. Cooling many hydrocarbons by 100°F increases their density by 5 percent.

- The pump's motor is somewhat oversized.

On this basis, cooling the liquid by 100°F will raise the pump's discharge pressure by 5 percent. If the pump is developing 1000 psig of differential pressure, cooling the liquid by 200°F will increase its discharge pressure to 1100 psig. If the discharge control valve is now opened, the pump's discharge pressure will drop back to 1000 psig, as the flow increases. The discharge pressure drops because the pump develops less feet of head at a higher flow rate.

To summarize, the specific gravity of the liquid has increased by 10 percent. The feet of head of liquid has decreased by 10 percent. Since:

$$\Delta P \approx (\text{specific gravity}) \times (\text{feet of head})$$

ΔP remains constant. But the lower required feet of head permits the pump to operate further out on its performance curve. This produces more volumetric (GPM) flow. It also requires more work to pump the greater flow. So we have to have sufficient excess amperage on the motor driver, to accommodate this maneuver. But that is the price we pay for expanding the pump's capacity, without increasing the impeller diameter. Of course, the larger-diameter impeller might still require more motor amperage.

24

Centrifugal Pumps: Driver Limits

Motors, Steam Turbines, and Gears

There are three types of limits on centrifugal pumps:

- *Impeller limit.* This was the subject of Chap. 23. Sometimes this is called a *pump limit.*

- *Driver limit.* This means the motor is tripping off on high amperage. Or, the turbine is running below its set speed.

- *Suction pressure limit.* Usually called lack of *net positive suction head* (NPSH), this is the subject of Chap. 25.

Motors

Pumps are driven by either fixed-speed or variable-speed machines. In this chapter, we consider all motor-driven pumps as fixed-speed machines. Certainly, there are some variable-speed motors, but these are few and far between, in most process plants.

A process motor in most American plants will be a three-phase motor, and operate at 3600 rpm or, less commonly, 1800 rpm. In European plants or facilities in South America once owned by European corporations, motors run at 3000 or 1500 rpm. At home in the United States, our motors are the two-phase variety, and operate at 1800 rpm.

There is an important difference between a two-phase and a three-phase motor. The three-phase motor can also be used as an electrical power generator. The two-phase motor, without modification, cannot. The difference is important, in that I once used this fact to avoid tripping off a critical process pump.

Helper turbine

Figure 24.1 is a sketch of a centrifugal pump, driven by a three-phase motor, with a turbine helper. This particular pump was charging a light gas oil stream to a high-pressure hydrocracker. The pump was operating quite close to its design conditions of

- ΔP = 2000 psig
- Flow = 17,000 bbl/day
- Specific gravity = 0.80
- Temperature = 400°F
- Speed = 3600 rpm

The operators were not running the turbine. The turbine was spinning, because it was coupled to the pump, but there was no motive steam to the turbine. The operators reported that the turbine was not needed, as the motor was pulling only 90 percent of its maximum amperage load. The question is, dear reader, whether the pump will run faster if the motive-steam flow is opened to the turbine. And the answer is, no. While the turbine will produce *shaft work,* and will help

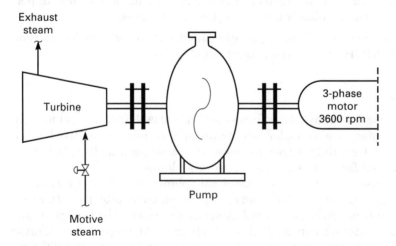

Figure 24.1 A pump driven by a motor with a turbine helper.

the motor spin the pump, the pump speed will remain unchanged. The extra shaft work from the turbine will just reduce the shaft work required from the motor shown in Fig. 24.1. The amp(erage) load on the motor will just be reduced.

But suppose the amp load on the motor drops to zero, by gradually increasing the steam to the turbine. Then, let's suppose we increase the motive steam to the turbine, by another notch. Will the pump, the motor, and the turbine (which are all coupled together and, hence, must run at the same speed) now run faster? The answer is no. But what will happen to the increment of shaft work, generated by the turbine?

This extra shaft work will be converted to electrical power. That is, the three-phase motor will turn into an electrical power generator. It will begin to export electricity into the grid. A two-phase motor does not have this sort of capability.

Another way of expressing this idea, is that the three-phase motor acts as a brake on the turbine. The motor can run at only 3600 rpm, as long as it is connected to the electrical grid. The only way we can get more flow from this pump without reducing the pump's ΔP is to raise the specific gravity of the liquid being pumped. If the liquid is cooled from 400 to 200°F, its specific gravity will increase by about 10 percent. This will allow us to get the same ΔP with about 10 percent less feet of head. This will allow the pump to run further out on its performance curve (see the last section of Chap. 23). The pump will now pump more liquid, and the flow of light gas oil to the hydrocracker could—and was—increased from 17,000 to 23,000 bbl/day.

Question: How does this affect the amp load on the motor driver? Well, the feet of head has gone down by about 10 percent, and the specific gravity has gone up by about 10 percent. Therefore, the horsepower or work per barrel of liquid pumped, remains constant. But the number of barrels has increased by 6000 bbl/day, or 35 percent. Then, the amp load on the motor driver would also increase by 35 percent. If I had not remembered this, prior to cooling the hydrocracker feed, the motor would have tripped off. As it was, I had the steam to the turbine increased, so as to decrease the motor amps to 60 percent of maximum, before cooling the gas oil. Omission of this detail, would have had embarrassing and possibly negative contractual consequences.

Selecting motor size

We noted in Chap. 23 that increasing the impeller diameter might require a larger motor for an existing pump. Suppose, then, when a new pump is purchased, we install a 100-hp motor sized for the maximum 10-in impeller, which the pump could accommodate. Actually, only an 8½-in impeller would be used in the pump initially. This 8½-in impeller requires only a 60-hp motor. If we install a 100-hp motor, what percent of the electrical power of the oversized motor is wasted?

Essentially, none. The amp load on the motor driver is determined by the work done by the pump, and not by the size of the motor. The amp load required by the motor, is proportional to

$$\frac{(\text{Pump head}) \times (\text{specific gravity}) \times (\text{liquid flow})}{(\text{Motor efficiency})}$$

There will be a small loss in motor efficiency, by using an oversized motor. As process operators and engineers, we can ignore this effect. It is good engineering practice to purchase new motors for the maximum-size impeller that can be installed in a pump.

Motor trip point

Factors which increase the amp load on a pump's motor driver are

- Increasing the flow, provided the pump is running on the flat portion of its curve.
- Increasing the specific gravity of the liquid pumped.
- Increasing the impeller size. A 10 percent increase in impeller size may increase the motor amp load by 33 percent.
- Motor winding deterioration, which happens as the motor ages, depending on how hot the motor runs.
- Dirt buildup on the motor's cooling-air fan guard screen.

Most operators forget to clean this screen. The dirty screen then restricts airflow to the motor, which then runs hotter. A hotter motor will pull more amps. I recently reduced the amp load on a motor by 3 percent by cleaning a badly fouled screen.

The amp meter on the motor breaker indicates the amps being pulled by the motor. Next to this meter should be a tag or a penciled number, indicating the full-limit amps (FLA) point for the motor. Above the FLA point, the motor should trip off. Operating the motor substantially above its FLA point will burn up the motor's windings.

This is not quite true. Whenever we start a pump, the starting torque required to get the pump spinning requires a surge of motor amps. To avoid tripping off the motor on high amps, there is a time delay built into the trip mechanism. This delay permits the amperage load to greatly exceed the FLA point, for up to 15 to 30 s. This is too brief a period in which to overheat the motor.

If we cannot open the discharge control valve of a centrifugal pump 100 percent before the FLA point is reached, then we say that the pump is *driver-limited.* This is a frustrating problem for plant operators, and clearly reflects poor design practice in undersizing the motor driver horsepower.

Motor bearings

If you think a motor is running rough because of bad bearings, have it uncoupled from the pump. Run the motor alone, and see if the vibrations persist. Motor bearings commonly fail because of lack of lubrication. Most motors are lubricated with a heavy grease, injected through a grease fitting, delivered via a grease gun. Overgreasing can burn up the motor bearings. You need to follow the manufacturer's recommendations. A typical (but not general) program for motor bearing lubrication, is three or four squirts from a grease gun, every 6 months. There is a device now available on the market called a "bearing buddy," which delivers a consistent grease supply pressure over an extended period of time. Motor bearing failures on one process facility disappeared when these bearing buddies were installed.

Steam Turbines

Turbine drives

A turbine-driven pump is said to be driver-limited when the governor speed-control valve is wide open. This speed-control valve, is usually called the "Woodward governor." It is not that easy to see if the governor is really wide open. A few simple methods to make this determination are

- Increase the pump's flow. If the turbine slows, the governor is wide open.
- Increase the set-point speed. If the turbine speed fails to come up, the governor was already wide open.
- Throttle on the steam supply to the governor. If the governor is already wide open, the turbine will slow down.

Even if the governor-steam inlet control valve is 100 percent open, it may still be possible to increase the supply of motive steam into the turbine. The position of the governor when it is wide open can often be reset to admit more steam flow. I once increased the flow through a crude tower-bottom pump by pushing hard against the base of the governor, and forced it to actually open to its maximum position. The pump speed came up 300 rpm, and the flow increased by about 15 percent.

Another possibility is to open the speed or hand valve, as described in Chap. 17, "Steam Turbines." This will introduce more steam into the turbine steam chest, and lower the steam pressure in the turbine case. Even though the governor-steam speed-control valve is wide open, the reduction in downstream steam pressure will increase the motive-steam flow, into the turbine.

Of course, increasing the motive-steam pressure will greatly increase the horsepower available to the turbine. First, more work can be extracted from each pound of the motive steam; and second, increasing the steam pressure will increase the pounds of steam flowing through the governor.

Reducing the turbine exhaust pressure will not significantly increase the steam flow to the turbine. However, reducing the turbine exhaust-steam pressure will increase the amount of work that may be extracted from each pound of steam. For example, 150-psig, 400°F steam is being used to drive a turbine. The exhaust steam is flowing into a 15-psig steam header. By venting of the steam to the atmosphere, the amount of work that can be extracted from each pound of steam will increase by roughly 30 percent.

Increasing size of steam nozzles

Many steam turbines do not have full-ported steam nozzles (see Chap. 17). The existing steam nozzles may then be exchanged for larger nozzles. An increase of nozzle diameter of 10 percent would increase the turbine horsepower by 20 percent.

Finally, the steam turbine's buckets can foul with hardness deposits from the steam. This reduces the turbine efficiency, and may prevent a pump from running at its rated speed. Injecting steam condensate into the steam supply can remove such deposits.

It is quite important not to operate a turbine-driven pump by throttling the steam flow to the turbine. Let's assume that the operators have set the turbine speed at 3500 rpm, by adjusting the steam inlet gate valve upstream of a malfunctioning governor. Suddenly, the discharge flow-control valve cuts back, and the pump's flow decreases from 2000 to 1200 GPM. The pump speed will then increase, because fewer pounds of liquid are being pumped, and less horsepower is required to spin the pump.

The pump speed rises to 3800 rpm. The trip speed has been set at 3750 rpm. The turbine's overspeed trip is unlatched, and the machine shuts down. The operator relatches the trip, but every time the flow is throttled back, the turbine overspeeds, and trips off.

Finally, the operator, in frustration, decides to wire up the trip so that it cannot unlatch. As an operator in a Gulf Coast refinery stated, "Norm, when the governor won't work on a turbine, it is necessary to wire up the trip." I thought he was joking—but he could not have been more serious.

Well, the operator did have a pump with an inoperable governor speed-control valve. He did wire the trip open. A few days later, the pump briefly lost suction, due to a slightly low level in the flash drum feeding the pump. The pump cavitated, and the flow was reduced. Reducing the flow to, or from, a centrifugal pump reduces the horsepower load on the driver. As the steam flow to the turbine driver was fixed, the turbine ran away. The excessive speed damaged the pump's bearing. The resulting vibration fractured the seal flush oil line to the pump. The flush oil ignited on the 700°F pump case. Everyone in the refinery has blamed me for the fire. Why, I cannot say. It is not my fault.

Gears

On occasion, pumps are not directly connected to either a motor or a turbine. There is an intervening gear, which can increase the pump's speed by multiplying the driver's speed. The gear is another source of possible misalignment and vibration. I have always considered such reduction or speed increaser gears to represent poor design practice, and an unnecessary complication as they often require their own lubrication system.

Centrifugal Pumps: Suction Pressure Limits

Cavitation and Net Positive Suction Head

The single most common operational problem in a process plant is loss of suction pressure to a centrifugal pump. If the suction pressure is too low, the discharge pressure and the discharge flow become erratically low. The suction pressure, while low, remains comparatively steady.

Cavitation and Net Positive Suction Head

The problem described above is called *cavitation*. A pump that is operating in a cavitation mode may also (but often does not) produce a sound similar to shaking a bucket full of nuts and bolts. A cavitating motor-driven pump always draws an erratically low-amperage flow. This is consistent with the erratically low flow rate.

(*Caution:* When a pump completely loses flow, it runs smooth, steady, and quiet. Its discharge pressure is stable. Its discharge flow is also stable and steady—a steady zero. This is not cavitation. The pump impeller is just spinning, to no particular purpose in the empty impeller case.)

Causes of cavitation

Let us assume liquid flows from an 8-in line into the suction of a centrifugal pump. The liquid enters the pump's impeller through a circular opening, called the "eye of the impeller," in the center of the impeller. Let us assume that this eye has a diameter of 2 in.

The velocity of the liquid increases by a factor of 16 (i.e., 8 in ÷ 2 in, squared). The kinetic energy of the liquid increases by a factor of 264 (i.e., 16, squared) But where does this large increase in kinetic energy come from? Answer: from the pressure, or feet of head, of the liquid itself.

The conversion of the pump's suction pressure to velocity in the eye of the impeller is called the *required net positive suction head* (NPSH). As the flow-control valve on the discharge of the pump shown in Fig. 25.1 is opened, the velocity of liquid in the eye of the impeller goes up. More of the pump's suction pressure, or feet of head, is converted to velocity, or kinetic energy. This means that the required NPSH of a pump increases as the volumetric flow through the pump increases.

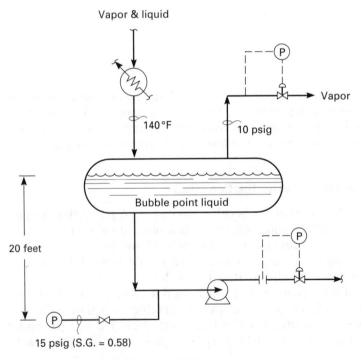

Figure 25.1   Available NPSH equals 20 ft.

The units of NPSH are feet of liquid head. The required NPSH of a pump is due primarily to the conversion of feet of head to velocity in the eye of the impeller.

The *available NPSH* to a pump, has the following definition:

- Physical pressure pump at suction

  Minus

- Vapor pressure of liquid at pump suction

When the required NPSH of a pump equals the NPSH available to the pump, the pump will cavitate.

Cavitation illustrated

Let us see what *cavitation* means in reference to the pump shown in Fig. 25.1. The liquid shown in the vessel is presumed to be in equilibrium with the vapor leaving the drum. This means that the liquid is at its bubble-point pressure and the vapor is at its dew-point temperature.

The vapor pressure of the liquid is then 24.7 psia (10 psig plus 14.7 psia of atmospheric pressure). The physical pressure at the suction of the pump is measured at 29.7 psia (15 psig plus 14.7 psia atmospheric pressure).

Then the physical pressure at the suction of the pump of 29.7 psia, minus the vapor pressure of the liquid at the suction of the pump of 24.7 psia, is 5 psia. This will be the NPSH available to the pump after we convert from psia to feet of liquid head.

The specific gravity of the liquid being pumped, as shown in Fig. 25.1 is 0.58 SG. To convert from psia to feet of liquid head, we have

$$5 \text{ psia} \times \frac{2.31}{0.58} = 20 \text{ ft of head}$$

The 20 ft matches the level of liquid in the drum above the suction line of the pump, shown in Fig. 25.1, and equals the available NPSH to the pump.

The required NPSH of the pump may be read from Fig. 25.2 (regardless of the SG of the liquid being pumped). It shows that at 250 GPM, the required NPSH of 20 ft, will equal the available NPSH of 20 ft. Therefore, at a flow rate of 250 GPM, the pump will cavitate. This calculation has neglected frictional losses in the suction line, which should be subtracted from the available NPSH.

Let's now assume that we wish to pump 300 GPM, not 250 GPM. If we open the flow-control valve shown in Fig. 25.1, the flow will momentarily increase. But, within a few seconds, the flow will become erratically low as the pump cavitates. The problem is that, according to Fig.

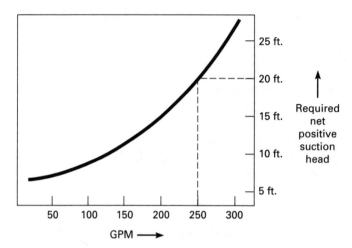

Figure 25.2   Required NPSH increases with flow.

25.2, we require an additional 6 ft of NPSH to increase the flow from 250 to 300 GPM.

One way of getting this extra suction pressure, or NPSH, is to raise the liquid level in the drum. For a liquid of 0.58 SG, with every 4 ft we raise the level in the drum, the suction pressure will increase by 1 psi, and the available NPSH will increase by 4 ft. But, unfortunately, the drum shown in Fig. 25.1 is already almost full.

Suppose we increase the pressure by partly closing the backpressure control valve. This will quickly increase the pressure in the drum from 10 to 15 psig. The pressure at the suction of the pump will also increase, from 15 to 20 psig. However, will this provide more NPSH to the pump?

Answer—no! Unfortunately, it is not only the pressure of the liquid at the pump that changes. The composition of the liquid will also be altered. As the pressure in the drum increases, additional lighter components dissolve in the liquid. The composition of the liquid then becomes lighter. The vapor pressure of the liquid will also increase by 5 psi. This must happen because the liquid in the drum, which is in equilibrium with the vapor, is at its bubble-point pressure.

Again, the available NPSH is the physical pressure at the suction of the pump, minus the vapor pressure of the liquid at the suction of the pump. If both pressures increase by 5 psi, the net gain in NPSH is zero.

Let us try again. Suppose we decrease the temperature of the drum, shown in Fig. 25.1, from 140 to 110°F. This will also cool the liquid flowing into the suction of the pump by 30°F. Will this colder liquid then provide more NPSH to the pump, by suppressing the flashing of the liquid flowing into the pump's impeller?

Answer—no! Unfortunately, it is not only the temperature of the liquid that changes. The composition of the liquid has also been altered. As the temperature in the drum decreases, additional lighter components dissolve in the liquid. The composition of the liquid becomes lighter. The liquid will now boil, not at 140°F, but at 110°F. This must happen because the liquid in the drum, which is in equilibrium with the vapor, is at its bubble-point temperature, which is now a cooler 110°F.

Let's try one more time. Suppose we tear off the insulation on the suction line and on the pump case and then spray cool water on the bare line. The temperature of the liquid in the drum, which is at 140°F, cools as it flows into the pump. By the time the liquid reaches the eye of impeller, it has cooled to 135°F. Will this slightly colder liquid provide more NPSH to the pump?

Answer—yes! But why? Well, the liquid is cooled by 5°F after it leaves the drum. The cooled liquid is not in equilibrium with the vapor in the drum. It has been *subcooled* by 5°F. This means that the bubble-point liquid has been cooled, without altering its composition. The vapor pressure of the liquid has been reduced. As can be seen in Fig. 25.3, subcooling this particular liquid by 5°F reduces its vapor pressure by about 2 psi. As the specific gravity of the liquid is 0.58, this is equivalent to an increase in the NPSH by 8 ft. Once again, our objective is to increase the flow from 250 to 300 GPM. Figure 25.2 tells us that the required NPSH increases from 20 to 26 ft. However, when we subcool the liquid by 5°F, the available NPSH increases from 20 to 28 ft. As the available NPSH now exceeds the required NPSH by 2 ft, the flow can be increased without risk of pump cavitation.

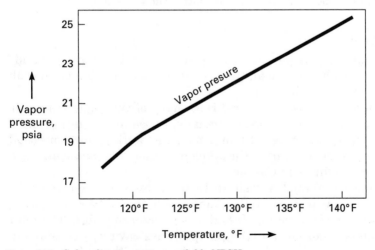

Figure 25.3 Subcooling increases available NPSH.

Starting NPSH requirement

The required NPSH read from the manufacturer's pump curve is called the *running NPSH*. However, when a pump is put on line, there is an additional type of NPSH requirement. This is called the *starting NPSH*. The initial velocity of the liquid in the suction of the pump is zero. After the pump is up and running, the velocity of the liquid in the suction of the pump might be 6 ft/s. This means that the liquid in the suction line has to be accelerated, which requires energy. This energy does not come from the pump; it must come from the liquid in the suction line itself. The only source of energy the liquid has is its pressure. This means that the pressure of the liquid is converted into kinetic energy. The kinetic energy accelerates the liquid from 0 to 6 ft/s.

This results in a temporary loss of pressure at the suction of the pump. This temporary loss of pressure is called the *starting NPSH requirement*. The more quickly the operator opens the discharge valve of a pump, the more rapidly the liquid accelerates in the suction line. This increases the starting NPSH required.

The longer the suction line and the larger the diameter of the line, the more mass has to be accelerated. This also increases the starting NPSH required. If the sum of the frictional loss in the suction line, plus the running NPSH, plus the starting NPSH, equals the available NPSH, then the pump will cavitate on start-up.

The experienced design engineer always allows for the starting NPSH requirement when determining the elevation of a vessel. It is hard to say whether to allow an extra 2 ft, or an extra 10 ft. It depends on the size of the suction line, and how carefully the operators are likely to be when starting the pump. But if the designer forgets this factor, then the plant operators are sure to notice the omission.

A hunting story

Gentle reader, thank you for reading so much of our book. Liz and I know you must be tired of all this technology. So let's change the subject. Let's talk about hunting.

How many of you are hunters? How many of you enjoy killing the innocent creatures of the forest? Actually, I used to go hunting once a year in Casper, Wyoming. At that time, I was working for Amoco Oil. So, once a year at the start of the antelope season, I found an excuse to visit Amoco's refinery in Casper.

I remember well my last visit, in 1980. My excuse to visit the refinery that year was to start up the pump shown in Fig. 25.4. This was a new, flashed crude-oil pump. It had never been run before. The refinery operators reported that the pump always cavitated on start-up. They had raised the liquid level in the drum to within a few inches of

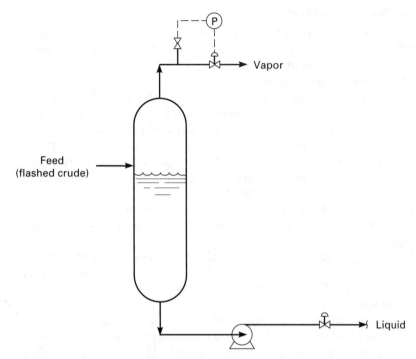

Figure 25.4 Overcoming a starting NPSH limit.

the inlet nozzle. A higher level would cause entrainment of black crude oil into the vapor outlet. The operators had then started the motor driving the pump. Next they very slowly opened the discharge valve.

Regardless of their efforts, the pump cavitated so badly that it could not be put in service. Some pumps are like that. The design engineer just never allowed any extra liquid elevation for the starting NPSH. But what could I do?

You see, guys, here is the problem. I cannot go hunting until I get the pump going. And I do not have much time. My friends are waiting for me at the plant gate. They're waiting in those great-big, high-wheel pickup trucks, designed for off-road use. They're blowing their horns and yelling for me to hurry up. It is a macho-type thing!

For my part, I cannot see the rush. I would just as soon stop at the gates of the city of Casper, Wyoming, and shoot the first antelope we came across. But not my friends. They want to go deep out into the back country to search for the "Ghost Antelope." And they are in a big hurry to get started. What, then, can I do to get the flashed crude pump up and running in just a few minutes? We shall see if you will just read on for the next few paragraphs.

Temporary increase in NPSH

This pump is presumed to run fine once it is running. The available NPSH is such that it exceeds the running NPSH. So how can I provide a temporary increase in the available NPSH, to satisfy the temporary starting NPSH requirement?

Answer: suddenly increase the pressure in the drum, by partly closing the backpressure-control valve shown in Fig. 25.4. This will instantly increase the pressure at the suction of the pump. It is true, as we said before, that raising the pressure in a drum does not increase the available NPSH, assuming that the vapor and liquid are at equilibrium. The idea of equilibrium assumes that the vapor is at its dew point and the liquid is at its bubble point.

As soon as the drum pressure is raised, the vapor composition in the drum is altered. The vapor composition becomes lighter. The vapor, though, is still at its dew point.

As soon as the drum pressure is raised, the liquid composition in the drum is altered. The liquid composition at the vapor-liquid interface becomes lighter. The liquid formed at the vapor-liquid interface is still at its bubble point.

When this lighter liquid works its way down to the suction of the pump, the beneficial effect achieved by raising the pressure in the drum is gone. The available steady-state NPSH will be exactly what it was before the drum pressure was raised. But this will take *time.* If the residence time of the liquid in the drum is 10 min, then it will take 10 min for the lighter liquid to reach the suction of the pump.

During this 10-min interval, the liquid flowing into the pump is the older, heavier composition. If I raise the pressure in the drum suddenly by 5 psi, then this instantly supplies about 20 ft of additional available NPSH—but only for a period of less than 10 min. During this period, I can crack open the pump discharge valve, push the motor START button, and then slowly accelerate the liquid in the suction line, by slowly opening the discharge valve. If I can do all this before the lighter liquid—formed at the vapor-liquid interface in the drum—reaches the eye of the pump's impeller, then I can start up the pump. Thus, even though we lacked the extra available NPSH to satisfy the pump's starting NPSH requirement, I could still go antelope hunting with my friends.

My pal, "one-shot" Bob Boening, did manage to shoot an antelope with a single round. As we approached the dying creature, it staggered to its feet and ran off. Bob and I then spent the rest of the day crawling and clambering through the ravines and rocky hills, south of Casper, searching for what did prove to be a "ghost antelope."

Why some pumps cavitate

Pumps cavitate for three reasons:

- They lack sufficient available NPSH to satisfy the conversion of pressure to velocity, in the eye of the impeller (running NPSH).
- They lack sufficient available NPSH to satisfy the conversion of pressure to acceleration, in the suction line, as the pump is started (starting NPSH).
- They lack sufficient available NPSH to overcome the frictional losses in the suction piping and the drain or draw nozzle.

It is positively my experience that the most common reason for pumps' cavitation is partial plugging of draw nozzles. This problem is illustrated in Fig. 25.5. This is the side draw-off from a fractionator. Slowly opening the pump's discharge control valve increases flow up to a point. Beyond this point, the pump's discharge pressure and discharge flow become erratically low. It is obvious, then, that the pump is cavitating.

The fluid being pumped is hot water. At the desired flow rate of 110 GPM, the manufacturer's pump curve shows that the pump requires 14 ft of NPSH. The elevation difference between the draw-off nozzle, and the suction of the pump is shown on Fig. 25.5, as 46 ft. We really ought to have plenty of running NPSH. But apparently, we do not.

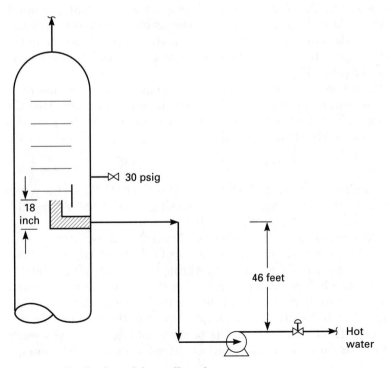

Figure 25.5 Partly plugged draw-off nozzle.

If I reduce the flow of water down to 100 GPM, the cavitation ceases. I now put a pressure gauge on the suction of the pump. Assuming that the suction line is full of 46 ft of water, what suction pressure would I expect to observe? Answer:

$$\frac{46 \text{ ft}}{2.3 \text{ ft/psi}} + 30 \text{ psig} = 50 \text{ psig}$$

The 2.3 ft/psi assumes that the specific gravity of water is 1.00.

But the observed pressure is not 50 psig. It is only 47 psig. I am missing 3 psig, or 7 ft of liquid:

$$(50 \text{ psig} - 47 \text{ psig}) \times 2.3 \text{ ft/psi} = 7 \text{ ft}$$

The most likely explanation for this *head loss* of 7 ft is frictional loss in the suction line. This reduces the available NPSH from 46 to 39 ft. But this is still a lot more available NPSH than the 14 ft of required NPSH needed to pump 110 GPM.

If I now open the discharge flow-control valve sufficient to increase the flow from 100 to 110 GPM, or by 10 percent, this would increase the frictional loss in the suction piping by about 21 percent, or about 0.5 psi, because ΔP varies with (flow)^2.

But this is not what I have observed. The suction pressure in Fig. 25.5 slips slowly down from 47 to 34 psig, at which point the pump begins to cavitate. What is happening? How could a 10 percent increase in the flow rate through the pump cause a 400 percent increase in the pressure drop in the suction line? What has happened to the lost 13 psig (i.e., 47 psig−36 psig)?

The boiling-point pressure of the water is equal to 30 psig (the pressure in the tower shown in Fig. 25.5); that is, we can assume that the water draw-off is at its bubble-point pressure. At 36 psig pump suction pressure, the available NPSH is then

$$(36 \text{ psig} - 30 \text{ psig}) \times 2.3 = 14 \text{ ft}$$

This matches the required NPSH, at a flow of 110 GPM, so the pump cavitates. But it still seems as if I am missing at least half of the 46 ft of liquid head to the pump. Where is it?

Well, dear reader, it no longer exists. Figure 25.6 illustrates the true situation. Let's say we are pumping 110 GPM from the pump discharge. But only 109 GPM can drain through the draw-off nozzle. We would then slowly lower the water level in the suction line. The water level would creep down, as would the pump's suction pressure. When the water level in the suction line dropped to 14 ft, the pump would cavitate or slip. The flow rate from the pump would drop, and the water level in the suction line to the pump would partially refill. The pump's

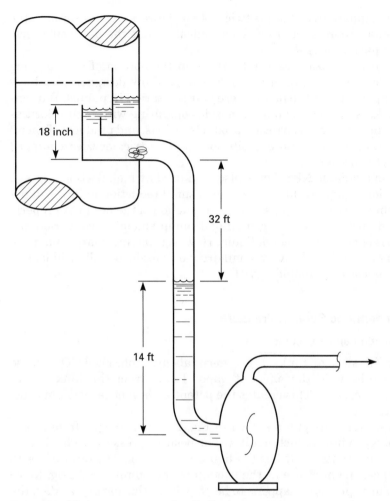

Figure 25.6 Most common cause of cavitation.

NPSH requirement would then be temporarily satisfied. Normal pump operation would be restored—but only for a moment.

Of course, it may simply be that the draw-off nozzle is undersized. To determine whether this is the case, calculate the velocity (in feet per second) V through the nozzle:

$$\Delta H = 0.34(V)^2$$

where ΔH is the hydraulic head in inches required to push 110 GPM of liquid through the draw-off nozzle. In this case, ΔH is found to be 9 in of water. Apparently, there is twice as much pressure loss through the nozzle than there should be. This indicates that the draw-off nozzle must be partly plugged.

Many draw nozzles, especially those in the bottom of vessels, plug because of the presence of *vortex breakers*. Many designers routinely add complex vortex breakers to prevent cavitation in pumps. But vortex breakers are needed only in nozzles operating with high velocities and low liquid levels. Corrosion products, debris, and products of chemical degradation can more easily foul and restrict nozzles equipped with vortex breakers.

Lack of available NPSH may also be caused by high frictional loss in the suction piping. If this is the case, a small reduction in flow will not noticeably increase the pressure at the suction of the pump. A properly designed suction line to a centrifugal pump should have a frictional head loss of only a few feet of liquid. However, having a large-diameter suction line, and a relatively small draw-off nozzle, usually will lead to excessive loss of available NPSH.

Subatmospheric Suction Pressure

Pump suction under vacuum

Liquids in storage tanks are almost always subcooled. This is so because otherwise the ambient vapor losses from the tank's vent, would be excessive. This creates the potential for a negative pump suction pressure.

For example, the pump shown in Fig. 25.7 is suffering from a partially plugged in-line suction filter. The positive pressure to the filter of 5 psig is due to the 15-ft head of liquid in the tank. The filter ΔP is 12 psi. Hence, the pressure at the suction of the pump is -7 psig. Since atmospheric pressure happens to be 15 psia on this particular day, the pressure at the suction of the pump is 8 psia.

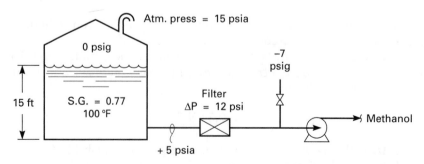

Figure 25.7 Pumping liquids with subatmospheric suction pressure.

The liquid being pumped is methanol, which has a vapor pressure, at the pumping temperature of 100°F, of 3 psi. The physical pressure at the suction of the pump (8 psia), minus the vapor pressure of the liquid at the suction of the pump (3 psi), equals 5 psi. To convert the 5 psi to feet of head:

$$\frac{(5 \text{ psi}) \times (2.3)}{0.77 \text{ (SG)}} = 15 \text{ ft}$$

The 15 ft of head is the available NPSH to this pump. Does this mean that pumps may have a substantial amount of available NPSH, even when their suction pressure is under a partial vacuum? Yes, if we are pumping a subcooled liquid. But this is quite common, because the liquid stored in an ordinary atmospheric-pressure storage tank is almost always well below its boiling point—that is, the liquid is subcooled.

Sump pumps

The most common type of pump used is the sump pump shown in Fig. 25.8. The vast majority of the pumps in the world are of this type. They are the sort of pumps used to pump water out of shallow wells and from irrigation ditches. In New Orleans, we use thousands of these pumps to push rainwater over the levees and into the mighty Mississippi River.

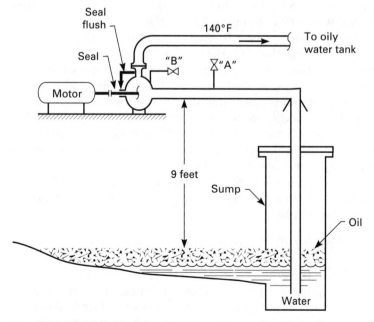

Figure 25.8 NPSH available to a sump pump.

Sump pumps can draw water up from levels as much as 30 ft below the pump's suction. But do such pumps require NPSH? Absolutely! All centrifugal pumps have some NPSH requirements. What, then, is the available NPSH to the sump pump shown in Fig. 25.8?

The physical pressure at the suction of the pump, is the pressure at point "A." The pump was lifting water from an oily-water sump. The water level in the sump was 9 ft below the centerline of the pump's inlet. Nine feet of water is equal to 4 psi (i.e., 9÷2.31). This means that the pressure at point "A" was −4 psig.

Atmospheric pressure on this particular day was 14 psia. The absolute pressure at the suction of the pump was then

$$14 \text{ psia} - 4 \text{ psig} = 10 \text{ psia}$$

The vapor pressure of water at 140°F pumping temperature is 3 psia. The suction pressure at the pump, minus the vapor pressure of water, at the suction of the pump is then 7 psi (i.e., 10 psia minus 3 psi).

The suction line itself, was a rather rough, old cast-iron pipe, with a frictional loss of 2 psi. This frictional loss, must be subtracted from the 7 psi just calculated. This leaves 5 psi available for us to convert to feet:

$$(5 \text{ psi}) \times (2.31 \text{ ft/psi}) = 11.5 \text{ ft}$$

This 11.5 ft is the NPSH available to the pump. The pump itself requires only 6 ft of NPSH to pump 1200 GPM of water. Hence, even though the pump's suction is 9 ft above the water in the sump, the available NPSH is twice the required NPSH.

A pump that is lifting very cold water (with a very low vapor pressure) through a smooth (low-frictional-loss) suction line, with a very small NPSH requirement, operating at sea level (where atmospheric pressures are high), can lift water to its suction by perhaps 30 ft. The pump shown in Fig. 25.8 can lift 1200 GPM of water by only 11.5 ft. When the water level in the sump drops to 11.5 ft below the centerline of the pump's impeller, the pump will cavitate.

Where, then, does the available NPSH to a sump pump really come from? It comes from atmospheric pressure. Atmospheric pressure, at sea level, is equivalent to

$$(14.7 \text{ psia}) \times (2.31 \text{ ft of water/psi}) = 34 \text{ ft of water}$$

In theory, this is the greatest height that water can be lifted from a well by a surface pump. In practice, this maximum lift height is 25 to 30 ft. Of course, water is pumped from bore-hole wells hundreds of feet deep. But this is done by submersible pumps, which are let down into the well.

Loss of prime

The only real difference between a sump pump and an ordinary centrifugal pump is that the sump pump is more difficult to *prime* than the ordinary pump.

To prime an ordinary pump, we simply crack open the vent valve on top of the pump's case. The pressure of the liquid in the suction line pushes out the gas or air trapped in the pump case to the atmosphere or to the plant's flare line header. Once the pump case is full of liquid, it is primed and ready to go.

The problem with priming the pump shown in Fig. 25.8 is that the water needed to fill the pump's case (i.e., to prime the pump) is below the pump itself.

Certainly, there are self-priming pumps. These pumps are designed to compress very, very small volumes of air, and to produce very, very small discharge pressures with the air. The circulation pump on my swimming pool is of this type. After running 15 min or so, it draws through itself a few cubic feet of air, before it can pick up suction. Most high-head, large-volume pumps must be primed with an external source of water. In Fig. 25.8, this is done by connecting a water hose to valve "A" and opening valve "B." Air is pushed out of the pump case by the pressure of water in the hose.

When the pump is in operation, the suction pressure of the pump is −4 psig. This negative pressure creates a potential problem. Air may be drawn into the suction of the pump through leaks in the suction line. One especially vulnerable area is the packing gland around the valve stem of the pump's suction valve. Another potential area of air in-leakage is around the pump's mechanical seal. Air drawn into either area will cause the pump to cavitate and lose its prime. The flow through the pump will then be lost. Playing a water hose over a leaking valve stem packing gland or over a bad seal can temporarily restore the pump to normal operation.

Self-flushed pumps

The pump shown in Fig. 25.8 is an illustration of an installation in Port Arthur, Texas. The difficulty with this pump was that it would lose its prime after being shut down for just a few moments. What is the problem?

Answer—there is an error in the design of this pump. The error is that this is a self-flushed pump. The mechanical seals on centrifugal pumps require a lubricant or seal flush material to keep the seal faces from touching and rubbing. In most pumps, this seal flush fluid comes from the discharge of the pump itself. Such pumps are called *self-flushed pumps.*

When a self-flushed pump is running, the space between the seal faces is filled with the seal flush fluid. When the pump is shut down, the space between the seal faces is filled with the fluid in the suction of the pump. However, if the pressure at the suction of the pump is below atmospheric pressure, then air is drawn through the seal faces and into the suction of the pump. This air displaces the water in the pump's case and, with time, causes the pump to lose its prime.

To fix this problem in Port Arthur, I connected an external source of seal flush water to the pump, from a nearby washwater station. Pumps which have subatmospheric suction pressures, and which are not in continuous service, should not be self-flushed pumps. They should have an external source of seal-flush material connected to the mechanical seal.

With the external source of seal flush, the pump no longer lost its prime when shut down. However, I then noticed an interesting phenomenon. The pump would continue to run for about an hour after it stopped raining. It produced a discharge pressure of about 100 psig. Then, quite suddenly, the discharge pressure would drop to 70 psig and the amp load on the motor driver would slip from 20 to 14 A for its last few minutes of operation.

The cause of this odd behavior is that the water level in the sump had dropped. A layer of oil with a specific gravity of 0.70 had been drawn into the suction of the pump. The feet of head developed by the pump had not changed. But a pump's discharge pressure and the amperage load on the motor driver is proportional to:

$$(\text{Density*}) \times (\text{feet of head})$$

---

*The density of the pumped fluid had dropped by 30 percent.

26

Separators: Vapor-Hydrocarbon-Water

Liquid Settling Rates

This chapter precedes the chapter on centrifugal compressors, for a sound reason. Damage to high-speed centrifugal compressors is most often associated with slugs of liquid entering the compressor from an upstream knockout drum. The other major use of knockout drums is to separate droplets of liquid from gas streams to be used for fuel or further processing. Separating droplets of water from liquid hydrocarbon is also discussed in this chapter.

Gravity Settling

The majority of the process vessels you see in your plant are gravity, vapor-liquid separators. Their main purpose is to settle out droplets of entrained liquid from the upflowing gas. Factors that affect the settling rate of these droplets are

- *Droplet size.*   Big droplets settle faster than little droplets. Contrary to that nonsense we were taught in school, cannon balls do drop faster then BBs—if one accounts for the resistance of air (BBs are small spheres of lead shot used in air guns).

- *Density of vapor.*   The less dense the vapor, the faster the droplet settling rate.

- *Density of liquid.* The more dense the liquid droplets, the faster the droplet settling rate.
- *Velocity of vapor.* The slower the vertical velocity of the vapor, the faster the settling rate.
- *Viscosity of the vapor.* The lower the vapor viscosity, the faster the settling rate.

All these ideas have been put into one equation, called *Stokes' law.* Nothing against Sir Frederick Stokes, but vapor viscosities are almost always so small that they do not affect settling rates. Also, we never know the particle size distribution of the droplets. There is a more fruitful way to look at the settling tendency of droplets of liquid in an upflowing vapor stream, as shown in Fig. 26.1. The method states

- The lifting force of the vapor, is proportional to the momentum of the vapor.
- The lifting force of the vapor is inversely proportional to the density of the liquid, which we will call D_L (lb/ft^3).

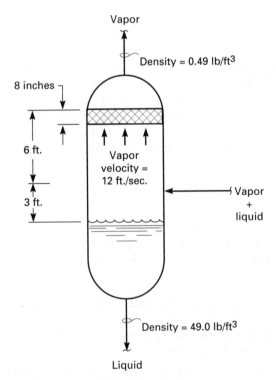

Figure 26.1   A vapor-liquid knockout drum.

I'll put this in equation form:

$$\text{Tendency to entrain droplets} = \frac{\text{momentum vapor}}{D_L} \qquad (26.1)$$

The momentum of the vapor equals

$$M \cdot Vg \qquad (26.2)$$

where M = mass of vapor, lb; and Vg = velocity of vapor, ft/s.

The mass flow, or weight of the vapor flow, per unit of vessel cross-sectional area is

$$M = D_V Vg \qquad (26.3)$$

where D_V = density of vapor, lb/ft^3.

Substituting Eqs. (26.2) and (26.3) into Eq. (26.1) gives us

Tendency to entrain

$$\text{droplets of Liquid} \quad = \frac{D_V}{D_L}VgVg = \frac{D_V}{D_L} \cdot Vg^2 \qquad (26.4)$$

Allow me now to call this tendency to entrain droplets K^2, where K is an empirically derived number, the *entrainment coefficient,* describing the tendency to entrain droplets of a randomly produced size distribution. We then have

$$K^2 = \frac{D_V}{D_L}Vg^2 \qquad (26.5)$$

or

$$Vg = K\left(\frac{D_L}{D_V}\right)^{1/2} \qquad (26.6)$$

Thousands of experiments have been run in the plant and the laboratory to determine K values for different levels of entrainment. Here are the results:

- $K = \leq 0.15$: very slight entrainment
- $K = 0.23$: normal to low entrainment
- $K = 0.35$: high entrainment
- $K = \geq 0.50$: very severe entrainment

I feel sure, dear reader, that you are objecting to my use of such terms as "slight" and "severe," as these are qualitative terms. Well, we do not know enough about entrainment to quantify it. Also, we have neglected such obviously important factors in the knockout drum, shown in Fig. 26.1 as:

- Vertical height between the inlet nozzle and the vapor outlet
- The clearance between the liquid level (or, more properly, the froth level) and the feed inlet
- The turbulence of the vapor and liquid, as they enter the drum
- The degree of dispersion of the liquid in the vapor

All I am saying, is don't take Eq. (26.6) or the tabulated K values as the ultimate truth. The calculated tendency of a vapor to entrain liquid is more akin to an educated guess, rather than a precise engineering calculation.

A simple K-value calculation might be helpful here. Let's use Eq. (26.6) and Fig. 26.1 to calculate the tendency to entrain droplets of liquid in the knockout (KO) drum shown:

$$12 = K \left(\frac{49.0}{0.49} \right)^{1/2}$$

$$K = \frac{12}{10} = 1.2$$

where the K value of 1.2 indicates that extremely severe entrainment would be certain from this KO drum. If we expected moderate entrainment, the volumetric gas flow to the vessel would have to be reduced by 80 percent, or the vessel diameter would have to more than double.

Normal vertical knockout drums are designed for a K value of about 0.20 to 0.25. If we are installing a KO drum ahead of a reciprocating compressor—and they really hate liquids in their feed—a K value of 0.14 might be selected. If we really do not care very much about entrainment, a K value of 0.4 might be selected. An example of this would be venting waste gas to the flare from a sour-water stripper reflux drum.

Demisters

Figure 26.1 shows an 8-in-thick demister below the top head. A *demister pad* resembles a giant Brillo pad without the soap. Many process plants have discarded demister pads lying around their scrap heaps, so you may have already seen what they look like.

The theory of operation of a demister is simple. Vapors and droplets of liquid strike the demister with a substantial velocity. The force of this impingement velocity causes the tiny droplets of liquid to coalesce into larger droplets. The heavier droplets drop out of the upflowing vapor.

For a KO drum with a demister to work properly, it apparently must have a K value of at least 0.15 to 0.20. I say "apparently" because in at least one service, I have noted an increase in entrainment from a vessel equipped with a demister at lower K values. This service was a sulfur plant final-effluent condenser. I suppose the vapors must strike the demister's fibers with some minimum force, to encourage droplets to coalesce.

Knockout drums equipped with demisters, according to what is written in other books, can tolerate K values 30 to 50 percent higher than can ordinary KO drums. More importantly, demisters definitely do scrub out much smaller droplets from a flowing gas stream than could settle out by gravity settling alone. The most common example of effective demister use is the removal of entrained boiler feedwater from steam generators. Boilers equipped with demisters produce steam of better quality; that is, the removal of the entrained droplets of water are removed and also hardness deposits are removed from the flowing steam. This is especially beneficial in keeping downstream superheater tubes and steam turbine rotor blades free of hardness deposits.

Demister failure

A rather large percentage of demisters I have encountered wind up on the scrap heap. Usually, this is due to plugging. If the flowing vapor is entraining corrosion products, coke, or other particulates, the demister will probably plug. The particulates will stick to the demister's fibers. Also, I have seen terrible demister plugging due to corrosion of the demister's fibers. Because of the large surface area of the demister, a relatively low corrosion rate will produce a large amount of corrosion deposits. It is much like that Brillo pad you left by the kitchen sink for a few weeks.

When a demister plugs, it increases the pressure drop of the vapor. But, the pressure drop cannot increase a lot, because the demister will break. Demister failure creates two problems:

- The dislodged sections of the demister pad are blown into downstream equipment, as into the suction of a centrifugal wet-gas compressor.

- The failed demister promotes high localized velocities. Vapor blows through the open areas of the vessel. The remaining sections of the demister pad, impede vapor flow. The resulting high, localized vapor velocities create more entrainment than we could have had without any demister.

Knockout drums handling gas with particulates are usually better off without a demister, unless there is a provision to backwash the demister. The demister must be made out of a corrosion-resistant material. In many refinery services, a high-molybdenum-content (high-moly) stainless steel is sufficient.

A little bit of entrainment is not always bad. For example, entraining a clean naphtha mist into the suction of a wet-gas centrifugal compressor is fine. The mist helps keep the rotor parts clean. If the steam from a boiler is to be used to reboil towers, entrainment is of no consequence.

Entrainment Due to Foam

A light, frothy foam is naturally more susceptible to entrainment than is a clear, settled liquid. Knockout drums handling foam, therefore, should be designed for a lower K value. Perhaps the K value that can be tolerated might be reduced by 25 to 35 percent, in the anticipation of such foam.

But foam may have a rather dramatic effect on entrainment, through a mechanism that is unrelated to vapor velocity. The problem I am referring to is high foam level. We already discussed this problem in Chap. 7, in the section devoted to level control. Figure 26.2 illustrates this problem.

The level shown in the gauge glass is 18 in below the inlet nozzle. This is okay. But inside the drum, we have foam. The average density of the foam plus liquid in the drum might be 0.60. The density of the clear liquid in the gauge glass is 0.80. Referring back to the logic in Chap. 7, the foam level in the tower is then 33 percent higher than the liquid level in the gauge glass.

Once this foam level rises above the feed inlet nozzle, the vapor in the feed blows the foam up the drum. A massive carryover of liquid (or foam) into the vapor line then results. I mentioned above that a small amount of mist carryover to a centrifugal compressor helped keep the rotor clean, and did no harm. Sudden slugs of liquid into the compressor inlet are another matter. They are likely to unbalance the rotor and cause substantial damage. If the compressor is a high-speed (10,000-rpm), multistage machine, it is likely to be wrecked by a slug of liquid. Even a small surge of liquid entering a reciprocating compressor can be very bad. The *intake valves* are almost certain to be damaged. Typically, the valve plate will break.

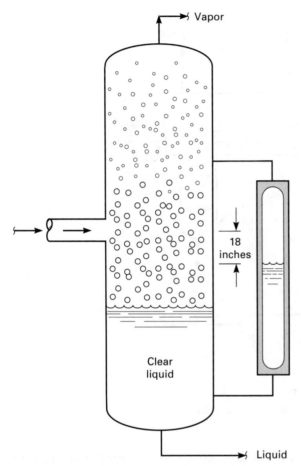

Figure 26.2 Carryover of foam due to a high froth level.

Water-Hydrocarbon Separations

Figure 26.3 shows a reflux drum serving a distillation column. We have a reflux drum because

- The drum provides a few minutes of holdup for the overhead product and reflux. This prevents the reflux pump from losing suction, should the amount of liquid discharging from the pump briefly exceed the amount of liquid draining from the condenser.

- The reflux drum separates liquid and wet gas by gravity settling. A horizontal vapor-liquid separator works in much the same way as the vertical KO drum.

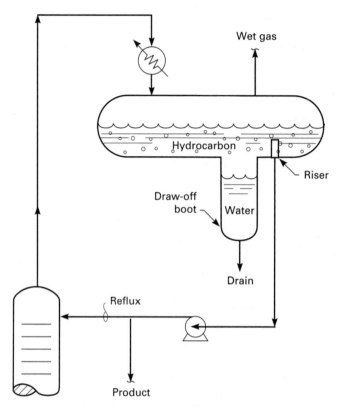

Figure 26.3 Water draw-off boot on a reflux drum.

- The horizontal reflux drum also separates hydrocarbon liquid from water.

Why do we usually want to separate water from the reflux stream, shown in Fig. 26.3? Some of the bad things that happen to the distillation tower, if water persistently entrains into the reflux are

- The distillation tray efficiency is reduced. The water may settle out on the hydrocarbon liquid on the tray. This reduces contact between the upflowing vapor and the downflowing, internal reflux.

- The tray may flood. Water and hydrocarbon mixing on the tray deck, stirred up by the flowing gas, creates an *emulsion*. The emulsion does not separate as readily as clear liquid from the gas. Premature downcomer backup, followed by tray deck flooding, result.

- Tray deck, downcomer, and vessel wall corrosion is increased. Water conducts electricity a thousand times better than do liquid hydrocarbons. Corrosion involves the transfer of electrons between steel and inorganic molecules, such as

- Hydrochloric acid to form ferric chlorides
- Hydrogen cyanide to form ferric cyanide
- Carbon dioxide to form iron carbonate
- Hydrogen sulfide to form iron bisulfide

The water acts as a highway for the electrons (the term *electrons* comes from the word *electricity*) to move between these potentially corrosive molecules and the vessel wall.

Water in reflux tends to get trapped in the tower, if the tower-bottom temperature is above the boiling point of water, at the tower's operating pressure. The water trickles down the tower and revaporizes off of the hot reboiler tubes. As the water may be saturated with corrosive salts and gases, reboiler tube corrosion can be rapid. In almost every petroleum refinery service, refluxing water is a quick route to reboiler tube leaks.

Water settling and viscosity

Water droplets settle out of a lighter liquid hydrocarbon phase because of gravity. The water is denser than the liquid hydrocarbons. Liquid droplets settle out of a lighter gas phase because of gravity. What is the difference? The difference is viscosity. We neglected the effect on the gas viscosity when calculating the K value in KO drums. Gas viscosities are almost always very low.

But liquid viscosities are extremely variable. At a constant viscosity, the settling rate of water in another liquid, due to gravity is proportional to

(Density of water) − (density of liquid in the reflux drum)

For water settling out of gasoline, in the reflux drum shown in Fig. 26.3, the settling rate is between ½ and 1 ft/min. If a droplet of water has to fall through a layer of hydrocarbon 3 ft deep, it would take about 3 to 6 min. If the hydrocarbon is less dense and less viscous than gasoline (like butane), the settling rate will be faster. If the hydrocarbon is more dense and more viscous than gasoline (like diesel oil), settling time will be longer.

The most neglected, but most important, feature of the reflux drum shown in Fig. 26.3 is the *riser,* a piece of pipe 4 to 12 in high, protruding from the bottom of the drum. You see, water first settles to the bottom of the drum. The water then runs along the bottom of the drum into the water *draw-off boot.*

Sometimes, corrosion products fill the bottom of the drum to a depth of 4 to 12 in. Sometimes holes corrode through the base of the riser, or the riser falls off. Either way, settled water can be drawn off the reflux drum into the suction of reflux pump.

Interface-level control

Water is drained off the boot to maintain an interface level between water and hydrocarbon in the boot. These interface level controllers, especially on reflux drum water draw-off boots, are typically set locally. The operator finds the visible interface level in the boot's gauge glass, and then sets the local interface level controller, to hold the level half way down the boot. Level taps in the lower portion of any vessel, tend to plug. This is especially true with the level taps in water draw-off boots. You need to blow out these taps on some frequent basis, because a high boot interface level can push water into the tower's reflux pump.

Electrically Accelerated Water Coalescing

Electrostatic precipitation

One really good method to speed up the settling rate of water is with electricity. Water is a *polar molecule,* meaning that one side of the water molecule is positively charged and the other side is negatively charged. It is this polar characteristic of water that makes it a good conductor of electricity. Thus, water is a highway for electrons. For example, we have the story of the operator who encountered a small electrical fire and, not having any water handy, decided to extinguish the fire by urinating on it. That was the last decision that operator ever made.

Hydrocarbons, on the other hand, are nonpolar molecules. They are not particularly affected by electricity. We can take advantage of this difference, between polar water and nonpolar hydrocarbons, to accelerate water settling.

An electric precipitator contains sets of electric plates. A high-voltage (20,000-V) electric current is applied across these plates, or electrodes. The droplets of water are electrically attracted to these plates, or grids. The water droplets coalesce into larger heavier droplets on the plates. They fall rapidly to the bottom of the electric precipitator vessel.

These precipitators may increase the settling rate of water by a factor of ~5. For example, settling droplets of sulfuric acid out of alkylate (a light hydrocarbon) by gravity alone might take an hour. A good electric precipitator can do a better job in just 10 min. Precipitators do not really consume much electricity. It is true that the electricity is applied at a very high voltage to speed up settling, but the amount of current drawn—or rather the amount of current that is *supposed* to be drawn—is nil.

If you see the amps on your precipitator creeping up, or spiking up, something is beginning to short-circuit the electric grids, or insulators. Most commonly, corrosion products are falling off the walls of the precipitator vessel. In my experience, this is the most common cause of precipitator failure.

Electric precipitators in mist removal

On my sulfuric acid production unit in Texas City, we had an electrostatic precipitator to remove a liquid sulfuric acid mist from a flowing gas stream. It worked in the same way as a precipitator in liquid-liquid service. However, the electrodes or grids were not parallel plates. As illustrated in Fig. 26.4, the grids were lead tubes and lead-coated wires.

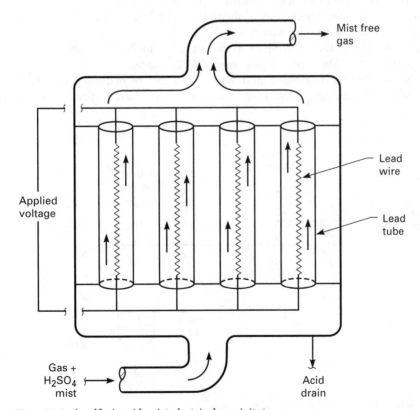

Figure 26.4 A sulfuric acid–mist electrical precipitator.

The gas flowed through about four hundred 8-in lead tubes, arranged in parallel. It was rather like a single-pass, shell-and-tube heat exchanger (see Chap. 21).

This electric sulfuric acid–mist precipitator worked fine, until one of the lead coated wires inside the tubes broke. Then the entire precipitator would arc. This means, the amps would jump up and down, as the single broken wire flapped around inside the tube. I do not know why, but the wires failed only on weekends. Then your fearless author would crawl inside the precipitator, and cut out the broken wire and plug the tube with a wooden plug.

Static Coalescers

A *coalescer* works in the same way as a demister, except that it is used to accelerate the removal of droplets of a heavier liquid from a flowing lighter liquid. An ordinary coalescer is shown in Fig. 26.5. This coalescer was used to remove entrained caustic from a flowing isobutane stream. The liquid isobutane would impact the coalescer pad at a velocity of 1 to 2 ft/min. The droplets of caustic, which have a higher surface tension than isobutane, would adhere to the surface of the coalescer fibers. As the caustic droplets grew bigger and heavier, they would drain down the fibers of the pad, and into the boot.

This particular coalescer pad was made out of ordinary straw. It would gradually plug and lose its effectiveness every few days. To

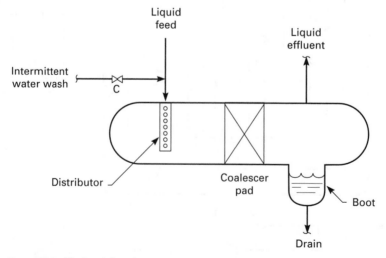

Figure 26.5 Horizontal coalescer.

restore its efficiency, we would open the intermittent water wash valve, shown in Fig. 26.5, for 15 min, to wash off the offending deposits.

Well, one day (this was in 1974), I decided that the use of straw in a modern industrial process plant was inappropriate. I ordered a new coalescer pad, constructed from state-of-the-art synthetic fibers. We replaced the sorry-looking straw pad with the really attractive, modern fiber pad—which did not coalesce nearly as well as the straw. The moral of this story is, of course, "If it ain't broke, don't fix it."

27

Gas Compression: The Basic Idea

The Second Law of Thermodynamics Made Easy

This chapter establishes the basis for the Second Law of Thermodynamics. It is not critical that you read this chapter to be able to understand the more practical chapters on compression that follow. But, for those readers who have technical training, wouldn't it be lovely to actually understand the basis for the Second Law of Thermodynamics. Wouldn't it be grand to really see the beauty and simplicity of the basis for the adiabatic compression work equation:

$$\text{Work} = P_1 V_1 \frac{K}{K-1} \left[\left(\frac{P_2}{P_1} \right)^{(K-1)/K} - 1 \right]$$

I have also written this chapter so that the nontechnical reader can easily comprehend the basis for this Second Law.

Relationship between Heat and Work

Robert Julius Mayer was a physician practicing near Bavaria in the 1840s. As part of his research into human metabolism, he decided to determine the equivalence between heat and work.

Heat means British thermal units, or the amount of fuel we have to burn to increase the temperature of a pound of water by one degree

Fahrenheit. *Work* means foot-pounds, or the amount of effort needed to raise a one pound brick by one foot.

The experiments that people like Dr. Mayer performed established the technical basis for the industrial revolution. Dr. Mayer himself laid the foundation for the main pillar supporting this technical basis. This was the science of thermodynamics. But in the nineteenth century, they had not coined the word *thermodynamics*. They called it "heat in motion."[1] The branch of science which we now call thermodynamics was developed by simply heating air under different conditions.

For example, let us pretend we are heating air with a wax candle. The air is confined inside a glass cylinder. We can assume that all the heat generated by burning the wax is absorbed by the air inside the cylinder. This is called an *adiabatic process*. I have shown a picture of the cylinder in Fig. 27.1.

The air in this cylinder is confined by a glass piston. The edges of the piston have been greased, so that the piston can glide without friction, up and down through the cylinder. But the edges of the piston itself have been so carefully ground that no air can slip between the greased piston and the walls of the cylinder. This means that the pounds of air contained in the cylinder below the piston will always be constant.

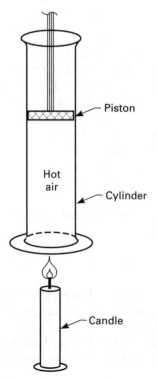

Figure 27.1 Measuring specific heats.

Dr. Mayer first wanted to determine how much wax he had to burn to heat the air inside the cylinder by 100°F. At this point in the experiment, he had to make a decision. Should he allow the expanding hot air to push the piston up? Or should he fix the position of the piston?

If the piston was kept in a fixed position, the pounds of air trapped inside the cylinder would continue to occupy a constant volume, as the air was heated. The pressure of the air would increase as it was heated, but the volume of air would remain constant.

If the piston was not kept in a fixed position, the pounds of air trapped inside the cylinder would stay at a constant pressure as the air was heated. The volume of the air would increase as it was heated, but the pressure of the air would remain constant.

Which is the correct way to conduct this experiment? Again, the objective of this experiment is to see how much wax has to be burned (which I will now call *heat*), to increase the temperature of a fixed weight of air by 100°F. Well, Dr. Mayer decided to conduct the experiment in both ways, to see if it made any difference. And this decision by Dr. Mayer was a turning point in human history.

Dr. Mayer already knew the weight of wax needed to heat 1 lb of water by 1°F. His British colleagues had previously determined this quantity, and had called it a *British thermal unit* (Btu).

He began by heating the air with the piston in a fixed position. Thus the volume of air heated was kept constant. The amount of heat (in Btu) needed to heat a fixed weight of air, under constant volume conditions, he called C_v. This is now called the *specific heat of air at constant volume*.

Next, Dr. Mayer heated the air, but allowed the piston to rise, as the hot air expanded. This kept the pressure in the cylinder just a little bit above atmospheric pressure. The Btu needed to heat a fixed weight of air, under constant pressure conditions, he called C_p.

Well, C_p turned out to be a lot higher than C_v. Dr. Mayer thought about this, and concluded that he had made a mistake in his experiment. The mistake was the weight of the piston. When he measured C_p, he had forgotten about the work needed to raise the heavy piston. Some of the heat generated by the burning wax was being converted to work to raise the heavy piston (work equals foot-pounds).

So Dr. Mayer repeated his experiment. He made the piston so light that its weight could be neglected. This helped, but still, C_p—now known as the specific heat of air at constant pressure—persisted in being about 40 percent larger than C_v. He reasoned that the expanding air must still be doing work and, therefore, converting some of the heat from the candle into work. But if the piston no longer had any significant weight, what sort of work could the expanding air be doing? Dr. Mayer's answer to this question changed history.

The expanding air was doing *compression work*. But what was being compressed? Not the air inside the cylinder, which, according to Dr. Mayer, was doing the work. This air was kept at a constant pressure. Certainly this air was not being compressed. No, dear reader, Dr. Mayer said that it was the air outside the cylinder that was being compressed. In other words, all the other molecules in the sea of air surrounding our planet were pushed a little closer together, by the expanding air in the cylinder.

Or, the air, which expands on heating, does work on its surrounding environment. But if we heat the air and don't allow it to expand, it cannot do any work on its environment. Let's just stop for a moment and give all these words some real teeth.

Let's say I allow air to expand, but I do not supply it with any extra heat from an outside source. The air is expanding because I am allowing it to push away a piston in a cylinder. What do you think happens to the temperature of this expanding air? Isn't it true, that whenever a gas (like air and natural gas) expands, it gets colder?

Compression Work ($C_p - C_v$)

The specific heat at constant pressure (i.e., C_p) is a measure of the amount of heat we put into the air, trapped inside the cylinder as shown in Fig. 27.1. Some of this heat is used to increase the temperature of the trapped air by 1°F. The rest of the heat goes into the work required to force the piston up and hence compressing the air surrounding the planet earth.

The specific heat at constant volume (i.e., C_v) is a measure of the amount of heat we put into the air, trapped inside the cylinder. All this heat goes to increasing the temperature of the trapped air by 1°F. None of the heat goes into compression work, because the piston remains fixed.

The difference between C_p and C_v is then *compression work:*

$$\text{Work} = C_p - C_v$$

How can we determine how much work is being done? There are two ways to calculate the amount of compression work that the piston is doing on the atmosphere of air surrounding the planet:

Method 1.   Take the pounds of wax burned when C_p was measured. Then, take the pounds of wax burned when C_v was measured. Find the difference between the two weights of wax. Burning one pound of wax generates about 18,000 Btu/lb. Let's say we burned 8 lb of wax, when measuring C_p. We burned 6 lb of wax, when measuring C_v. The amount of heat that has then gone into work is

$$(8 \text{ lb} - 6 \text{ lb}) \times 18{,}000 = 36{,}000 \text{ Btu}$$

or

$$C_p - C_v = \text{work}$$

We currently know that one Btu is equal to 740 ft-lb worth of work. Therefore the work performed by the expanding air equals

$$740 \times 36{,}000 = 26{,}700{,}000 \text{ ft/lb}$$

If none of this work was wasted, we could use it to lift a 1000-lb rock up to the top of Mount Everest (which is about 26,000 ft high). But in the 1840s, Robert Mayer did not know the conversion factor of heat equivalent to work of 740 ft-lb per Btu. It had not been determined yet, because Dr. Mayer was the one who first discovered it. So Dr. Mayer had to use the following method.

Method 2.   Let's say the diameter of the piston in Fig. 27.1 was 40 ft. The area of such a piston is 1256 ft². The burning wax is causing the weightless piston to be pushed up by 100 ft by the air, as it is heated and expands. The piston is being pushed up against an atmospheric pressure of 14.7 psia. Let us remember that there are 144 in² in a square foot. So that I can say that atmospheric pressure is actually

$$14.7 \times 144 = 2117 \text{ lb/ft}^2$$

The total force of the atmospheric pressure pressing down on my piston is then

$$2117 \text{ lb/ft}^2 \times 1256 \text{ ft}^2 = 267{,}000 \text{ lb of force}$$

Well, work equals force times distance. The piston is traveling a distance of 100 ft. Therefore, the work done by the expanding air, is

$$100 \text{ ft} \times 267{,}000 \text{ lb} = 26{,}700{,}000 \text{ ft-lb}$$

$$\text{Distance} \times \text{force} = \text{work}$$

Dr. Mayer used our second method. He knew that the heat of combustion of 2 lb of wax was 36,000 Btu. He divided

$$\frac{26{,}700{,}000 \text{ ft-lb}}{36{,}000 \text{ Btu}} = 740 \frac{\text{ft-lb}}{\text{Btu}}$$

to obtain the heat equivalent of work. It would be impossible to design an industrial process unit without knowing this fact.

The difference $C_p - C_v$ is proportional to the amount of work the piston could perform when supplied with a total amount of heat, proportional to C_p. Then the ratio

$$\frac{C_p - C_v}{C_p} = \frac{C_p}{C_p} - \frac{C_v}{C_p} = 1 - \frac{C_v}{C_p}$$

is equal to the fraction of useful work we could recover from a total heat input proportional to C_p.

The term C_p/C_v is usually called K (the ratio of the specific heats). If I substituted K into the preceding equation, I would obtain

$$1 - \frac{1}{K} = \frac{K}{K} - \frac{1}{K} = \frac{K-1}{K}$$

Does that look familiar? It really ought to, if you have any type of engineering training. Remember the formula for compression work, given at the start of this chapter, and in our thermodynamics textbooks:

$$\text{Work} = P_1 V_1 \frac{K}{K-1}\left[\left(\frac{P_2}{P_1}\right)^{(K-1)/K} - 1\right]$$

where P_1 = suction pressure, psia
P_2 = discharge pressure, psia
V_1 = suction volume, ft^3
$K = Cp/Cv$

There is another, nonmathematical, way to think about the difference between C_p and C_v. To extract work from any process requires an energy input, like burning wax. Some of this energy will be extracted as work ($C_p - C_v$). But most of this energy input will wind up as heat (C_v).

This is the nasty, but inescapable, lesson learned from the Second Law of Thermodynamics, which is derived from Julius Mayer's work. A few examples will suffice to close this chapter:

- A modern car engine converts about 12 to 15 percent of the energy in the gasoline to shaft horsepower extracted from the engine.
- The most efficient engine ever made is the Rolls-Royce high-bypass jet engine that we see on commercial aircraft. This engine converts about 39 percent of the energy in the jet fuel (which is kerosene) to thrust.
- A modern power station converts about 30 percent of the energy in the fuel burned to exported electrical energy.
- The ratio of ($C_p - C_v)/C_p$ for air is about 30 percent.

Reference

1. J. Tyndall, *Heat a Mode of Motion*, 6th ed., Longmans, London, U.K., 1880.

28

Centrifugal Compressors and Surge

Overamping the Motor Driver

Have you ever heard a 12,000-hp, 9000-rpm, multistage centrifugal compressor go into *surge?* The periodic, deep-throated roar, emitted by the surging compressor, is just plain scary. Machines, quite obviously, are not intended to make such sounds. But what causes surge?

Another question: What happens to the amperage load on a motor-driven centrifugal compressor, when the molecular weight of the gas increases? I ask this question in the following context:

- The compressor is a fixed-speed machine, as shown in Fig. 28.1.

- The suction pressure P_1 is constant.

- The discharge pressure P_2 is constant.

- The number of moles of gas compressed or the *standard cubic feet per hour* (SCFH) is constant.

- The suction temperature is constant.

We ought to be able to answer this question with Robert Mayer's equation—also called the *Second Law of Thermodynamics* (see Chap. 27), which states that motor amperage (or electrical work) is proportional to

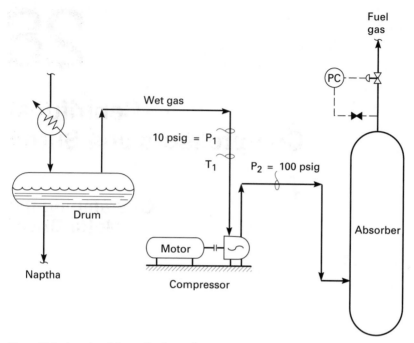

Figure 28.1 A motor-driven, fixed-speed, centrifugal compressor.

$$\text{Compression work} = NT_1 \frac{K}{K-1}\left[\left(\frac{P_2}{P_1}\right)^{(K-1)/K}-1\right]$$

where N = number of moles, a constant
  T_1 = suction temperature, a constant
  P_2 = discharge pressure, a constant
  P_1 = suction pressure, a constant
  K = ratio of the specific heats, C_p/C_v

We will assume that over the ranges of molecular weights that we are working with the ratio of the specific heats K is constant. This is not quite true, but this approximation will not invalidate the following statement: *According to the Second Law of Thermodynamics, as the molecular weight of the gas compressed increases, the amperage (amp) load on the motor should remain constant.*

The only problem with this statement is that it contradicts reality. When we actually increase the molecular weight of a gas, the amp load on the centrifugal wet gas compressor shown in Fig. 28.1, does increase. This seems to contradict the Second Law of Thermodynamics. But the Second Law has never been shown to be wrong. So we have a conflict. Our experience tells us that the amp load on the motor must increase as the molecular weight of the gas increases. But the Second Law of Thermodynamics tells us that the amp load on the motor must remain the same as the molecular weight of the gas increases.

The resolution of this conflict between theory and practice, and the question "What causes surge?" will require the rest of this chapter to answer.

Centrifugal Compression and Surge

Mechanically, what is surge?

What is actually happening inside a compressor when it begins to make that surging sound? Let us refer to Fig. 28.2. When a compressor starts to surge, the gas flows backward through the *rotating assembly* (i.e., the *rotor*). This reversal of flow pushes the rotor backward. The rotor slides backward along its radial bearings. The radial bearings support the weight of the rotor.

The end of the rotor's shaft now slams into the thrust bearing. The thrust bearing constrains the axial (i.e., horizontal), movement of the rotor. Each time you hear the compressor surge, the rotor is making one round trip across its radial bearings. Each time the rotor surges, the force of the end of the shaft impacting the thrust bearing causes the thrust bearing to deform. As the thrust bearing deforms, the axial movement of the rotor increases. The spinning wheels of the rotor come closer and closer to the stationary elements (called the *labyrinth seals*) of the compressor, which are fixed inside the compressor case.

When a spinning wheel (with a wheel tip velocity of perhaps 600 miles an hour) touches a stationary element, the compressor internals

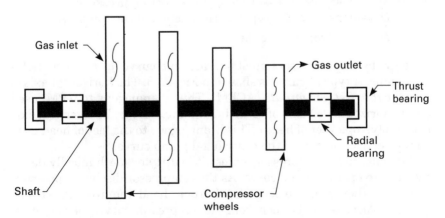

Figure 28.2 Rotating assembly for a centrifugal compressor.

are wrecked. Pieces of the wheel have been know to tear through the compressor case, and kill operators. Older (1960s), lower-speed compressors seem to withstand the destructive forces of surge better than do newer, higher-speed models.

How centrifugal compressors work

Centrifugal compressors and centrifugal pumps work on the same principle. If you have neglected to read Chap. 23, "Centrifugal Pumps: Fundamentals of Operation," this would be a good time to read it. Both centrifugal compressors and centrifugal pumps are *dynamic machines,* meaning that they convert velocity into feet of head.

The gas enters the compressor's rotor through the large wheel shown in Fig. 28.2. The purpose of this wheel is to increase the velocity or kinetic energy of the gas. After the high-velocity gas escapes from the vanes in the wheel, the gas enters the stationary elements fixed to the inner wall of the compressor case. This is called the *stator.* Inside the stator the velocity or kinetic energy of the gas is converted to polytropic feet of head, or potential energy.

Brave reader, do not be afraid of the term *polytropic feet of head.* It really has the same simple meaning as described in Chap. 23, except the term polytropic *feet of head* means feet of head for a compressible fluid that is changing temperature.

To convert from polytropic feet of head to ΔP, which is really what us process people are interested in, we use the following very rough approximation:

$$\Delta P \approx D_V \cdot H_p \tag{28.1}$$

where ΔP = discharge pressure minus the suction pressure
D_V = density of the vapor at the suction of the compressor
H_p = polytropic feet of head

Centrifugal compressors must operate on a curve, just like centrifugal pumps. A typical curve is shown in Fig. 28.3. The horizontal axis is *actual cubic feet per minute* (ACFM). This is analogous to GPM, used on the horizontal (x) axis of centrifugal pump curves. The vertical axis is H_p (polytropic feet of head). This is analogous to the feet of head used on the vertical (y) axis of the centrifugal pump curve.

The centrifugal compressor, unless it is dirty or mechanically defective, has to operate on its curve. As the compressor discharge pressure increases, then Hp, the feet of polytropic head required, must also increase. Also, as can be seen from the compressor curve, the volume of gas compressed (ACFM) must decrease. When the volume of gas drops

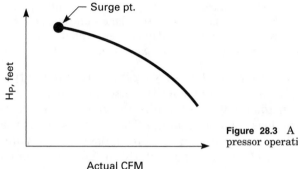

Figure 28.3 A centrifugal compressor operating curve.

below a critical flow, the compressor will be backed up to its surge point.

Aerodynamic stall

In my younger days, I used to try to meet women on airplanes. Finding myself seated next to an interesting lady, I would ask, "Have you ever wondered what makes this plane fly?" With this opening gambit, I would then explain:

> This sketch (Fig. 28.4)) is a cross section of the wing. Because of the shape of the wing, the air has to travel a longer distance across the top of the wing than underneath the wing. This means that the velocity of the air as it travels across the top of the wing is greater than the velocity of the air as it travels underneath the wing. The energy to increase the velocity, or kinetic energy of the air as it flows across the top of the wing, does not come from the plane's engine. This energy to accelerate the air comes from the air itself; that is, the increase in the kinetic energy of the air flowing across the top of the wing comes from the barometric pressure of the air.
>
> It follows, then, that the pressure on top of the wing (shown in Fig. 28.4) is less than the pressure underneath the wing. This difference in pressure,

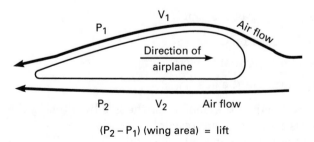

$$(P_2 - P_1) \text{ (wing area)} = \text{lift}$$

Figure 28.4   Aerodynamic lift.

multiplied by area of the wing, is called *lift*.

As the plane's air speed is reduced, its ability to maintain a lift equal to its weight is reduced. At some reduced speed, the plane's lift then becomes insufficient to keep it flying. The aircraft undergoes aerodynamic stall. The plane falls out of the sky, crashes, and all the passengers are killed.

At this point, the young lady whom I was trying to impress would typically pick up a magazine and ignore me for the rest of the journey.

Surge is quite similar to aerodynamic stall. Of course, when a compressor surges, its rotor does not stop spinning. The rotor is spun by the motor. But when the flow of gas through the rotor falls below a certain rate, the forward velocity of the gas stops. With no flow, there is no velocity to convert to feet of head. Then, the ΔP developed by the compressor falls to zero.

The discharge pressure of the compressor shown in Fig. 28.1 is 100 psig and its suction pressure is 10 psig. The gas flow, therefore, travels backward. The reverse gas flow pushes the rotor backward, and slams it up against its thrust bearing. The suction pressure of the compressor increases and its discharge pressure decreases. Temporarily, the ΔP required to push the gas from the wet-gas drum and into the absorber shown in Fig. 28.1 is reduced. The polytropic head requirement is thus also temporarily reduced. The compressor may then run out on its performance curve, as it moves a greater ACFM volume, and move away from surge. But in so doing, it lowers its own suction pressure, raises its own discharge pressure, and creates the conditions for the next destructive surge.

Required ΔP

Movement of the gas from the wet gas drum into the absorber requires a certain ΔP. According to Eq. (28.1)

$$\Delta P \approx (\text{vapor density}) \times (\text{polytropic head})$$

We can increase ΔP by either of the following options: raise the density of the vapor or raise the feet of polytropic head, developed by the compressor.

To raise the density of the gas, we could

- Raise the compressor's suction pressure
- Increase the molecular weight of the gas
- Decrease the temperature of the gas

Obviously, we cannot change the density of the gas by altering the mechanical characteristics of the compressor.

To raise H_p, the feet of polytropic head, we could

- Increase the number of wheels on the rotor shown in Fig. 28.2

- Increase the diameter of the wheels
- Increase the speed of the rotor

Obviously, we cannot change the feet of head developed by the compressor by altering the physical properties of the gas compressed.

It is true, then, that vapor density and feet of head are not related. But if the product of the two numbers do not result in sufficient ΔP to push the gas from the drum into the absorber, then the gas flow will stop. It will stop and then reverse its direction of flow. And that is what causes surge.

Too much polytropic head

You might conclude from my description of surge that the engineer needs to be cautious when designing a new compressor so that it will not surge. For example, let's assume that Jane has to issue the specifications for a new wet-gas centrifugal compressor. She checks with John, the unit engineer, for the proper molecular weight of the gas. John tells Jane that the molecular weight of the gas is normally 30, but it can be as low as 24—that is, the density of the gas can, on occasion, be 20 percent lower than normal.

Jane concludes that the lower-density gas will require more feet of polytropic head to develop the required ΔP. To avoid the possibility of surge, she decides to increase the number of wheels on the compressor from five to six. While Jane has used good engineering judgment, she has made a serious error. It turns out that John should not have been trusted. The actual molecular weight of the gas turns out not to be 24 or 30, but 36. The gas is 50 percent more dense than Jane's design specifications.

Poor Jane! The compressor's motor driver now trips off, on high amps! In her efforts to avoid surge, she has run afoul of the real-world fact, that the motor amps required to drive a centrifugal compressor are approximately proportional to the molecular weight of the gas—in apparent contradiction to the Second Law of Thermodynamics.

I hope that you can now see the intimate relationship between surge in a centrifugal compressor and the amperage load on the motor driver used to drive the compressor. Now, let's see if I cannot prove that the Second Law of Thermodynamics is quite in harmony with our practical experience.

Effect of molecular weight on ΔP

Let us refer again to Fig. 28.1. Suddenly, there is an increase in the molecular weight of the wet gas. This causes the density of the gas to increase. This results in an increase of the compressor ΔP. As the compressor ΔP increases, the compressor's suction pressure decreases.

Why? Well, if the discharge pressure is kept constant by the absorber backpressure control valve, then a bigger ΔP must drag down the suction pressure. The reduced suction pressure increases the suction volume (ACFM) of gas flowing to the compressor. Why? Because a lower-pressure gas occupies a larger volume.

As the ACFM increases, we run out to the right on the compressor curve, shown in Fig. 28.3. As we move away from the surge point, the polytropic feet of head decreases. As the polytropic feet of head is reduced, the compressor ΔP comes partially back down to its initial value, until a new equilibrium is established. But because the initial disturbance of the equilibrium—the increased molecular weight— moved us away from surge, the new equilibrium will be established farther away from surge than the initial equilibrium. Not only will the new equilibrium be established farther away from surge, but the pressure in the wet gas drum will also wind up lower than the initial pressure in the drum.

Let's now assume that there is a sudden decrease in the molecular weight of the wet gas. This results in a decrease in the gas density. The ΔP developed by the compressor goes down. As a consequence, the compressor's suction pressure rises. This reduces the ACFM volume of gas flowing into the compressor. As the ACFM decreases, we back up on the compressor curve toward the surge point. As we move closer to surge, the polytropic feet of head developed increases. The compressor ΔP comes partly back up to its initial value, until a new equilibrium is established. But because the initial disturbance—the decreased molecular weight—moved us toward surge, the new equilibrium will be established closer to surge than the initial equilibrium. Also, the pressure in the wet-gas drum will wind up higher than the initial equilibrium pressure in that drum.

Compressor Efficiency

Maintaining a constant suction drum pressure

In the two examples we have just discussed, you should have noticed, that P_1, the compressor suction pressure, was not being held constant. In fact, when the molecular weight of the gas increased, P_1, the suction pressure, went down. Let me now rewrite Robert Mayer's equation, which I presented at the start of this chapter:

$$\text{Amps} \approx N\left[\left(\frac{P_2}{P_1}\right)^{0.23} - 1\right] \qquad (28.2)$$

This simplified version assumes that K (the C_p/C_v ratio) is a constant of 1.3; that T_1, the suction temperature, is constant; and that the compression work is proportional to the motor amps (N is the number of moles of gas flowing into the compressor). From Eq. (28.2), it is obvious that a reduction in the suction pressure P_1 will cause the motor's amp load to increase.

But I said at the start of this chapter, that I assumed that P_1 was to remain constant. Well, I made an impossible assumption. Because of the dynamic nature of the centrifugal compressor, it is simply impossible to permit the molecular weight of a gas to vary, and then state that none of the following parameters may change:

- Suction pressure, P_1
- Discharge pressure, P_2
- Number of moles, N
- Speed of the compressor

As the molecular weight goes up, P_1 goes down, the compression ratio ($P2/P1$) increases, and so does the work needed to drive the compressor. And the apparent contradiction between Dr. Mayer's equation and our plant experience is resolved.

But, suppose we must maintain a constant pressure in the wet-gas drum. The pressure in this drum may be controlling the pressure in an upstream distillation column. To hold a constant pressure in the drum, we will have to resort to spillback suction pressure control, illustrated in Fig. 28.5.

Again, let's assume that there is a sudden increase in the molecular weight of the gas. Again, the gas density increases. The ΔP developed by the compressor rises. The pressure in the wet-gas drum drops. The spillback pressure-control (PC) valve starts to open. The number of moles N flowing to the compressor increases. The ACFM also increases. The compressor now runs out on its operating curve (Fig. 28.3) to the right, away from surge. The polytropic feet of head H_p drops. The ΔP is brought back down to its initial value. This keeps both the compressor suction pressure and the pressure in the wet-gas drum constant.

Great. But what about Dr. Mayer? What about Eq. (28.2)? It is true that the compression ratio ($P2/P1$) has been held constant. But as the molecular weight has increased, the number of moles N has also increased. Why? Because, the spillback valve has opened. And as N increases, so does the amperage drawn by the motor.

Why, then, in ordinary process plant practice, do we see an increase in the amps on a motor driving a centrifugal compressor as the gas becomes heavier? Does it take more work to compress a mole of propane [44 MW (molecular weight)] than it does to compress a mole of methane (16 MW)? Certainly not. It's just that compressing a heavier

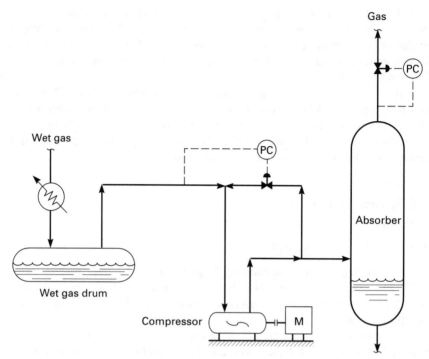

Figure 28.5 Spillback suction pressure control.

gas forces the spillback to open, to prevent the pressure from falling in the wet-gas drum. This extra gas recirculating through the compressor is the real factor that increases the amp load on the motor driver.

Variable-speed driver

Wouldn't it be more simple to just slow down the compressor as the molecular weight increases? After all, it seems as if we need less feet of head when the gas density increases. As we slow the compressor by 10 percent, the feet of polytropic head would drop by 20 percent. Fine. But 99 percent of motor-driven compressors are fixed-speed machines.

Does this mean that we would be better off driving a large centrifugal compressor with a variable-speed driver? Perhaps with a steam turbine or gas-fired turbine. You bet! Especially when the molecular weight is highly variable.

What about Jane?

I hope you all understand the mistake Jane made. She trusted John. But if she had it all to do over again, what could Jane have done differently? She has to drive the compressor by a fixed-speed motor. The

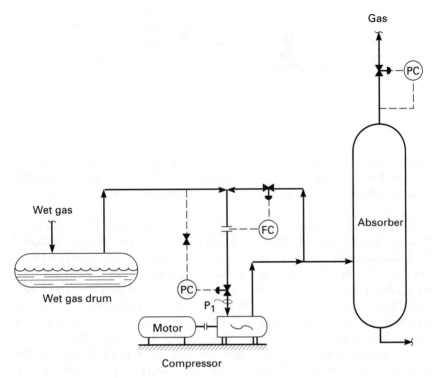

Figure 28.6 Suction throttling pressure control.

gas molecular weight is going to be unpredictable. The pressure in the wet-gas drum has to be kept constant.

Figure 28.6 is a partial answer to Jane's predicament. It is a suction throttle valve. Use of the suction throttle valve will partly reduce the increase in electrical work (amps), due to an increase in the molecular weight of the gas.

If the molecular weight of the gas increases, the gas density will increase. The ΔP will increase. The pressure in the wet-gas drum will drop. The new suction throttle PC valve will start to close. This will restore the pressure in the wet-gas drum, without increasing the flow of gas through the spillback valve. But what happens to P_1? How does closing the suction throttle valve affect the actual compressor suction pressure?

Gentle reader, take a break. You are up to the hardest part of this book. Have a Coke, and we will continue.

Damming the River Yeo

The River Yeo is a gentle stream that meanders through western England. In the seventeenth century, the river was used to move coal-

Figure 28.7   Dam on a river does not affect level downstream.

laden barges. Except near the town of Yeovil, the river became too shallow for deep draft barges to negotiate. So a dam was proposed. This would raise the height of water upstream of the dam, as shown in Fig. 28.7.

The farmers downstream of the dam were concerned that the river flowing through their fields would run dry. The engineer who designed the dam assured them that this would not happen. Certainly, at first the flow of water downstream of the dam would be just a trickle. But after a few months, when a new equilibrium was established, the water would begin overflowing the dam at the same rate that water formerly flowed down the river. The level of the river downstream of the dam would then be the same as it had been before the dam existed.

The suction throttle valve, shown in Fig. 28.6, is analogous to the dam on the River Yeo. When we close this valve, the pressure in the wet-gas drum will increase. But the pressure at P_1, the compressor suction pressure, is completely unaffected by the movement of the suction throttle valve.

For example, let's say that the molecular weight of the gas increases by 20 percent. The ΔP, developed by the compressor, would then increase from 90 to 108 psi. If the absorber pressure is fixed, then the wet-gas drum pressure would drop by 18 psi. To prevent this excessive decrease in the drum pressure, we throttle on the suction PC control valve. The pressure drop through this valve would increase by 18 psi. But the pressure at P_1 will drop by 18 psi, just as if we never moved the suction throttle valve.

Controlling the pressure in the drum by suction throttling allows the compression ratio $P2/P1$ to increase in response to an increase in the molecular weight of the gas. As the compression ratio increases, according to Eq. (28.2), the amperage on the motor driver also increases. But not by a lot.

According to Eq. (28.2), the compression ratio is raised to a small fractional exponent (viz., 0.23). So even if the compression ratio goes up a lot, the amp load on the motor driver will increase by very little.

Of course, we may also control the wet-gas drum pressure very nicely by the spillback PC valve, shown in Fig. 28.5. But this mode of control causes N, the number of moles, in Eq. (28.2), to rapidly increase. And the motor amps will increase just as fast as the number of moles.

The useful rule of thumb is then

- When the molecular weight rises, the motor amps must increase.
- If the molecular weight rises by 20 percent and we control the drum pressure by spillback control, the motor amps will increase by about 20 percent.
- If the molecular weight rises by 20 percent and we control the drum pressure by suction throttle control, the motor amps will increase by about 10 percent.

Saving electricity

Forget about Jane. Forget about John. At the moment, there is nothing wrong with the compressor shown in Fig. 28.6. We are operating far away from the surge point, and the motor is pulling only 70 percent of its maximum amp load.

Both the suction throttle pressure control (PC) valve and the spill-back [flow-control (FC)] valve shown in Fig. 28.6 are in a nice operating position. All we wish to do is save electricity.

This is done by reducing the set point on the FC valve. As we do this

- The FC valve partly closes.
- The number of moles compressed N is reduced.
- The volume of gas compressed (ACFM) goes down.
- According to the compressor curve shown in Fig. 28.3, the feet of polytropic head H_p increases as the ACFM decreases.
- The increased H_p increases the ΔP developed by the compressor.
- This reduces the pressure in the wet-gas drum.
- The PC valve shown in Fig. 28.6 now begins to close, to restore the pressure in the wet-gas drum.

The net result of these changes is to reduce N and increase $P2/P1$. According to Eq. (28.2), this should result in a reduction in the amp load of the motor driver. A rough rule of thumb is that for every 10 percent decrease in N, the total number of moles compressed, the amp load on the motor driver will fall by 5 percent.

But, as we reduce N, we are moving toward the surge point in Fig. 28.3. If we try to save too much electricity, by forcing the FC to close too much, then the centrifugal compressor may be forced into surge. That is why we call the FC spillback valve the *antisurge valve*.

Once again, please notice how the problem of surge and the amp load on the motor driver, are two ideas that interact together. It is kind of like solving two equations with two unknowns. The two objectives we are trying to optimize are the motor amps and avoidance of surge. The two handles we have on the problem are the suction throttle valve and the spillback valve. And just like solving the two equations, we have to optimize the position of the two valves, simultaneously.

Suction temperature

Throughout our discussion of motor-driven, constant-speed centrifugal compressors, I have assumed that the suction temperature was constant. But let's refer back to Fig. 28.1. Assume that the temperature in the drum increases. How will this affect the pressure in the drum?

If the increased temperature T_1 increases the density of the gas, then the ΔP developed by the compressor will go up, and the drum pressure will decline. Let's assume that the feed to the drum is a mixture of

- Ethane
- Propane
- Butane
- Pentane
- Gasoline

As the drum temperature increases, the heavier components in the liquid phase are vaporized into the vapor phase. This increases the molecular weight of the vapors and hence the vapor density.

Now, let's assume that the feed to the drum is a mixture of hydrogen and heavy mineral oil.

As the drum temperature increases, the molecular weight of the gas remains at 2, because the mineral oil is too heavy to vaporize. The density of the vapor goes down, because the vapor is hotter. The drum pressure then increases.

Sulfur plant air blower

Here is a problem that came up on a sulfur recovery facility in Punto Fijo, Venezuela. The combustion air blower, shown in Fig. 28.8, was a fixed-speed, motor-driven centrifugal machine. The air intake filters were severely fouled. They had a pressure drop of about 8 in H_2O. The atmospheric vent valve, used to control the discharge pressure at a constant 12 psig, was 50 percent open. The unit engineer had been asked to calculate the incentive in electrical power savings that would result from cleaning the filters.

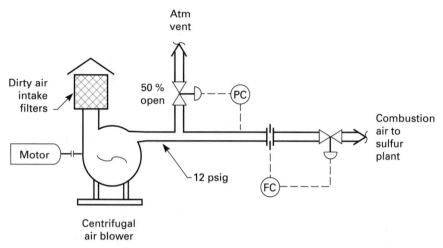

Figure 28.8 Filter plugging affects motor amps.

Cleaning the filters would have raised the suction pressure to the blower. This would have reduced the blower's compression ratio. But, because the blower is a fixed-speed centrifugal compressor, the amount of air compressed N, would have increased. The atmospheric vent discharge valve would have been forced open. The greater volume of air compressed would have caused the compressor to work harder, even though the compression ratio ($P2/P1$) was lower. Finally, cleaning the filters would have increased the amps drawn by the motor.

To prove my point, I slid a piece of plywood across one filter. The PC vent valve, shown in Fig. 28.8, began to close, and the amps drawn by the motor driver went down!

(*Caution:* The equations presented in this chapter for polytropic head and compression work have been simplified for clarity. They cannot be used for rigorous engineering calculations.)

29

Reciprocating Compressors

Positive Displacement and Carnot Cycle; Use of Indicator Card

Centrifugal compressors are dynamic machines. They convert velocity, imparted to the gas by a spinning wheel, to polytropic feet of head. The feet of head, multiplied by the density of the gas, equals the pressure boost produced by the centrifugal compressor.

A *reciprocating compressor* is a direct volume-reduction machine. The gas is simply squeezed out of a cylinder by a piston and pushed into the discharge line. The molecular weight of the gas does not influence the suction or discharge pressure of the compressor. The gas density does not influence the compressor performance or the work required by the driver.

The reciprocating compressor is a positive-displacement compressor. It is cheaper to purchase and install than a centrifugal compressor. Also—in theory—far more efficient (90 percent) than a centrifugal compressor (70 percent). Certainly, reciprocating compressors are more simple to understand and engineer than centrifugal machines. Best of all, they are not subject to surge.

There are only two real problems with reciprocating compressors: pulsation and mechanical reliability. But these problems are so intractable that, for most industrial applications, centrifugal compressors are preferred, except when dealing with low-molecular-weight gas. A low-molecular-weight gas, like hydrogen, has a low density. Let's say that a compressor must develop a large differential pressure or ΔP. Recalling Eq. (28.1) from the previous chapter:

$$\Delta P = (\text{gas density}) \times H_p$$

where H_p = polytropic feet of head.

A centrifugal compressor developing a lot of ΔP with a low gas density needs to produce a lot of polytropic feet of head. This means that the centrifugal compressor must

- Be a high-speed machine
- Have a lot of wheels or stages
- Have large-diameter wheels

The most cost-effective way to produce a high polytropic head is to increase the number of wheels on the rotor. But the longer the rotating assembly, the more difficult it is to properly balance the rotor. Especially for high-speed machines, rotors which become unbalanced are subject to destructive vibration. Therefore, for low-molecular-weight gas services (less than 10), it is not uncommon to use reciprocating compressors.

Reciprocating compressors are also favored over centrifugal machines for temporary installations such as gas field well-head service. Also, when low initial capital investment is favored over long-term maintenance costs, reciprocating compressors are often used.

Theory of Reciprocating Compressor Operation

Figure 29.1 is a simplified sketch of a cylinder of a reciprocating compressor. The cylinder is shown as a single-acting cylinder. Typically cylinders are *double-acting,* meaning that there are valves on both ends of the cylinder and that the piston is compressing gas, in turn, on both ends of the cylinder.

The far end of the cylinder is called the *head end.* The end of the cylinder nearest the central shaft is called the *crank end.* I have shown the valves only on the head end, to simplify my description of the compressor's operation.

There are four distinct steps in the compression cycle of the cylinder, shown in Fig. 29.1: compression, discharge, expansion, and intake.

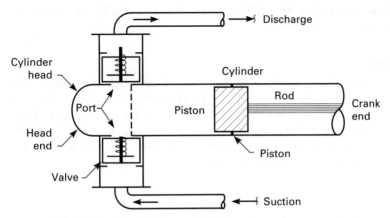

Figure 29.1 Reciprocating compressor cylinder.

Compression

The initial position of the piston is as far from the cylinder head as possible. This is the start of the compression stroke. This position is called *top dead center.* The piston now moves toward the cylinder head. The pressure of the gas inside the cylinder increases.

Discharge

At some point, as the piston approaches the cylinder head, the gas pressure inside the cylinder exceeds the pressure in the discharge line. The discharge valve now springs open, and gas is pushed out of the cylinder into the discharge line.

Both the discharge valve and the intake or suction valve are simply spring-loaded check valves. The discharge valve permits gas only to be forced out of the cylinder. The intake valve permits gas only to enter the cylinder.

The piston continues its travel toward the cylinder head. At some point it stops, and reverses its direction. This point is called *bottom dead center,* indicated by the dotted line in Fig. 29.1. Of course, bottom dead center cannot coincide with the end of the cylinder. The piston would have to travel past the *valve ports* for this to occur. If the piston travels past the discharge port, the compressed gas could not be pushed out of the cylinder into the discharge line. So bottom dead center must line up with the crank-end edge of the valve ports, as shown in Fig. 29.1.

This means that a substantial volume of gas remains trapped between the cylinder head and the piston, before the travel of the piston is reversed. This volume of gas, called the *starting volumetric clearance,* determines the *volumetric efficiency* of the reciprocating compressor. We will discuss this later.

Expansion

As soon as the piston reverses its direction of travel, the pressure of the gas inside the cylinder drops. The gas pressure drops below the discharge line pressure, and the spring-loaded discharge valve slams shut. The piston continues its travel toward the crank end of the cylinder. The pressure of the gas inside the cylinder continues to fall, but the suction or intake valve remains closed.

Intake

At some point, as the piston approaches the crank end, the gas pressure inside the cylinder falls below the pressure in the suction line. The suction, or intake, valve now springs open, and gas is drawn out of the suction line and into the cylinder. This portion of the intake stroke continues until the piston returns to top dead center.

The Carnot Cycle

The piston has now completed its cycle. I have drawn a picture of this cycle in Fig. 29.2. I have plotted the pressure inside the cylinder against

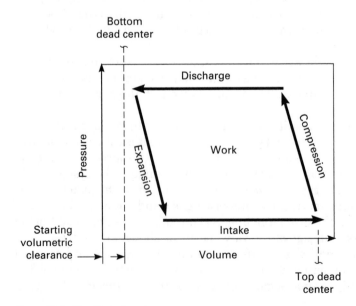

Figure 29.2 The Carnot cycle.

the volume of the gas inside the cylinder. The "gas inside the cylinder" refers to the gas between the piston and the cylinder head.

Beginning at top dead center:

1. The gas is compressed up to the discharge pressure.

2. The gas is pushed out into the discharge line, at a pressure equal to the discharge-line pressure.

3. The piston reaches bottom dead center, and reverses its direction.

4. The gas left in the end of the cylinder expands—and depressurizes—until the suction-line pressure is reached.

5. New gas is taken into the cylinder until the piston returns to top dead center.

At any point in this cycle, there is a pressure inside the cylinder that corresponds to the volume of gas inside the cylinder. That is all the *Carnot cycle* chart, drawn in Fig. 29.2, means.

Professor Nicolas L. S. Carnot, in the late nineteenth century, realized that the area inside the plot of pressure vs. volume represented the work needed to compress gas in a reciprocating compressor. In other words, the change of pressure, multiplied by the change in volume, is equal to the work done by the piston on the gas. Professor Carnot called this *PV* (pressure vs. volume) work. He then used calculus to sum up the area inside the lines shown in Fig. 29.2. The total area is now called *ideal compression work.*

What's wrong with Carnot's theory? I have nothing against Carnot personally. I do not hold his French nationality against him. I do not dislike French people, even though Liz's purse was stolen in Paris. But I do have several problems with the Carnot cycle:

- Valve leakage
- Spring tension
- Pulsation

The Carnot cycle plot represents ideal compression work. But we in the process industry have to worry about actual compression work, and the loss of compression efficiency caused by these three problems.

We would like to understand the actual compressor cycle, rather than the idealized Carnot cycle. We would like to see an actual plot of the pressure inside the cylinder, compared to the volume inside the cylinder. And there is a way to do this.

The Indicator Card

The pressure inside the cylinder can be measured with a *pressure transducer,* a device that converts a pressure into an electrical output. The pressure transducer screws into the end of the cylinder head. There is a screwed plug in the cylinder head for this purpose.

The volume of gas inside the cylinder is a function of the piston position. The piston position is a function of the crank-shaft position. The crank-shaft position can be measured by a magnetic pickup attached to the crank shaft. Anyone who has had an automobile spark-plug firing timing adjusted electronically is familiar with this method of determining piston position.

The output from the magnetic pickup and the output from the pressure transducer are connected to an oscilloscope. After a computer conversion of the data, the resulting pressure/volume plot is printed out. This plot is called an *indicator card.* The indicator card shown in Fig. 29.3 was generated from a 4000-hp natural-gas compressor in Hebronville, Texas.

The solid line is the indicator-card plot. The dotted line is the Carnot or ideal compression work cycle which I have drawn myself. The piston position, shown on the horizontal axis, is proportional to the volume of gas inside the cylinder.

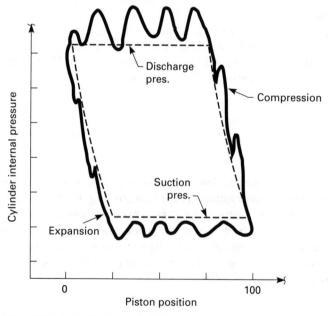

Figure 29.3 An indicator card plot.

The area enclosed by the solid line is the total or actual compression work. The area enclosed by the dotted line is ideal or useful compression work. The area between the dotted line and the solid line represent compression work lost to heat. The area inside the dotted line, divided by the area inside the solid line, is called *adiabatic compressor efficiency*.

Pulsation losses

Take a closer look at the discharge portion of the indicator card shown in Fig. 29.3. Note that the cylinder internal pressure rises well above the discharge-line pressure, before gas is pushed out of the cylinder and into the discharge line. This is a consequence on the spring tension of the discharge valves.

In order to force these valves open, a certain amount of extra pressure in the cylinder is needed. The extra pressure overcomes the spring tension. Once the valve is open, gas flows out of the cylinder, into the discharge line, and the cylinder pressure falls. But, as shown on the indicator card, the discharge valve apparently opens and closes five times during one cycle.

This multiple opening of the discharge valve is called *valve pulsation* or "valve flutter." The peaks on the indicator card, caused by this pulsation, are a source of considerable energy waste. The amplitude (i.e., the height of the peaks) is a function of the valve spring tension. The frequency of the peaks (i.e., the number of peaks) is a function of the speed of the compressor and the geometry of the suction and discharge piping.

We can reduce the amplitude (height) of the peaks by using weaker springs in the discharge valve. I tried this at the Hebronville compressor station, and it did work. The frequency (number) of peaks cannot be altered by any currently available cost-effective method.

When we reduce the spring tension, we prevent the valves from closing as tightly as they did with stronger springs. The discharge valves are now more prone to leakage. The effect of discharge-valve leakage can be seen in the expansion portion of the cycle, shown in Fig. 29.3.

The sudden jumps in pressure are caused by high-pressure gas blowing back from the discharge line into the cylinder. If the discharge-valve leakage gets bad enough, the amount of gas compressed may approach zero. While the amount of actual or total compression work may still be quite high, the amount of ideal or useful compression work may be essentially zero.

Using the indicator card

Is it then a good idea to make the discharge-valve springs weaker? Weaker springs cut down on pulsation losses, but also increase valve leakage losses. If we consult the indicator-card plot, we can see that the pulsation losses (for the discharge portion of the cycle) greatly exceed the valve leakage losses (for the expansion portion of the cycle). It therefore seems that weaker springs probably are a good bet, to promote greater compressor efficiency.

I could repeat the same story for the intake or suction valves. Leakage of the intake valves would appear as peaks on the compression portion of the cycle. Pulsation losses for the intake valves would appear as peaks on the indicator card during the suction or intake portion of the cycle.

Other problems that can be identified by use of the indicator card are

- Piston ring leakage
- Excessive valve velocity losses
- Breakup of the valve plates

The indicator card is the only real way to monitor reciprocating-compressor performance. Typically, the equipment and personnel to generate the card can be obtained from a local company specializing in this service. Often, the indicator card is referred to as a *beta scan plot*.

Volumetric Compressor Efficiency

Reciprocating compressors have two sorts of compressor efficiency:

- Adiabatic
- Volumetric

So far, we have limited our discussion to adiabatic compression efficiency. This sort of inefficiency downgrades work to heat. For a given compression ratio, the temperature rise of the gas as it flows through the compressor may be excessive, thus indicating a low adiabatic compression efficiency. Both centrifugal and reciprocating compressors suffer from this common problem, which is the subject of Chap. 30.

Volumetric efficiency applies only to reciprocating compressors. A reduction in volumetric efficiency reduces the gas flow through the compressor. A reduction in volumetric efficiency need not reduce the adiabatic compression efficiency.

A reduction in volumetric efficiency reduces the work required from the driver. For reciprocating compressors, we intentionally reduce the volumetric efficiency, to reduce the load on the driver. Sometimes this is done to save energy; sometimes this is done to prevent the motor driver from tripping off on high amps.

Unloaders

In the plant, we use the term *unloading* to indicate various ways of reducing the volumetric efficiency of a reciprocating compressor. There are two sorts of unloaders:

- Valve disablers.
- Head-end clearance pocket adjusters.

The proper way to reduce the volumetric efficiency is to increase the starting volumetric clearance. This is done with an adjustable unloading pocket, as shown in Fig. 29.4. This device, also called the *head-end unloader,* works by increasing the starting volumetric clearance. This is known as the volume of gas trapped between the cylinder head and the piston, when the piston position is at bottom dead center, as explained in Fig. 29.2.

Turning the wheel at the back end of the cylinder counterclockwise, pulls back a large internal plug in the head. Now, when the piston starts to withdraw toward the crank end of the cylinder, there is more gas left inside the cylinder to expand. The greater the volume of gas inside the cylinder when the piston is at bottom dead center, the closer the piston is to top dead center before the intake valve opens. The delay in the opening of the intake valve reduces the amount of gas drawn into the cylinder. This reduces the number of moles of gas compressed by the piston. Compression work also diminishes and the driver horsepower or amp load drops.

Valve disablers

The head-end unloader has a great advantage in that it unloads the driver without increasing the amount of work required, per mole of gas

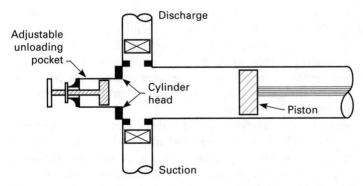

Figure 29.4 An unloading pocket reduces engine load and volumetric capacity.

compressed. The disadvantages of the adjustable clearance pocket or head-end unloader, shown in Fig. 29.4 are

- It is an expensive added feature.
- It cannot be used on the crank end of the cylinder.

In practice, the head-end unloader can reduce compression work required per cylinder only by roughly 30 percent.

The second way to unload a reciprocating compressor is with *valve disablers*. Most of the unloaders you have on your compressors are likely of this inferior type. They typically consist of steel fingers, which are pressed down through the valve-cap assembly. The fingers prevent the valve plate from moving. These valve disablers are far less costly than an adjustable clearance pocket. They may be used on both the crank and head ends of the cylinder. They can reduce the gas flow through a cylinder to zero.

Unfortunately, valve disablers have a detrimental effect on the adiabatic compressor efficiency. This means that, even though no gas may be moving through the crank end of a cylinder, the piston is still doing work on the gas inside the crank end of the cylinder. If you would like proof, place your hand on the valve cap on such a disabled cylinder. The high temperature you will feel is wasted compression work going to useless heat. I have measured in the field that, after a cylinder end is completely disabled, it is still converting 20 percent of the former compression work to heat.

The other problem with valve disablers is poor mechanical reliability. They get stuck inside the valve cap, especially in dirty-gas service. Once stuck, the valve—and really the entire end of the cylinder—is out of service, until the next overhaul of the reciprocating compressor.

Valve failure

The weak points of a reciprocating compressor are the intake and discharge valves. Valves fail because the movable valve plate or the springs break. The cause of failure is usually liquid entering the compressor. Reciprocating compressors do not like liquids. A small, well-dispersed liquid flow to a centrifugal compressor will help keep its rotor clean and does no damage. The same amount of liquid will tear up the valves on a reciprocating machine.

High temperature also hastens spring failure and plate cracking of the discharge valves. The cause of high discharge-valve temperature, is primarily valve leakage. Leaking valves cause a percentage of the gas flow to be recompressed by the piston. This extra compression work can result in some very hot discharge valves.

Valve leakage is caused by a combination of pulsation and by fouling deposits. Sulfur, sulfur salts, ammonium chloride, and other volatile compounds may all sublime inside the valve assemblies. The word *sublime* means to change directly from a vapor into a solid. Amine salts, polymers, oxygen, diolefins, water, and products of corrosion also contribute to valve fouling. These deposits accumulate between the valve plate and the fixed valve slotted face. They prevent proper seating of the movable valve plate. This results in valve leakage. Centrifugal compressors are far more tolerant of fouling then are reciprocating compressors.

Rod Loading

The piston rod used in a reciprocating-compressor cylinder can fail if overstressed. The stress experienced by the piston rod is a direct function of the ΔP developed by the cylinder. That is because on one side of the piston, we have the suction pressure and on the other side of the piston, we have the discharge pressure. The ΔP times the area of the piston is therefore the force acting on the piston. Excessive force will break the piston rod.

The high-discharge-temperature trip you have on your reciprocating compressor is intended to prevent the piston rod from being overstressed. The higher the ΔP, the greater the temperature rise of the compressed gas. The compressor manufacturer has calculated the expected temperature rise of the gas for the maximum ΔP to which the piston should be exposed. The high-temperature trip is not intended to protect mechanical components from excessive wear and tear, due to high temperature. It is there to protect the machine from the very serious consequences of piston-rod failure.

Do not reset this high-temperature trip above the manufacturer's specifications, without prior consultation with the equipment manufacturer. Ladies and gentlemen, you can imagine how I became so smart on this subject.

30

Compressor Efficiency

Let's assume that we are driving a centrifugal compressor with a constant-speed electric motor. We are compressing natural gas, coming right off a thousand wellheads, in Laredo, Texas. This is not a good idea. There is entrained brine (salty water) in the gas. The brine will dry out inside the compressor case, due to the heat of compression. The resulting salts will deposit on and inside the wheels or stages of the compressor's rotor.

The compressor efficiency will be adversely affected. As a consequence:

- The flow of gas compressed will be reduced.

- The discharge temperature of the compressor will increase.

- The amp load on the electric motor driver will go down.

Why, though, does it take less work to drive the compressor, when its rotor wheels are encrusted with salt? It is true that it takes somewhat more work to compress a mole of gas with a fouled rotor. But the fouled rotor also compresses a lot fewer moles of gas. Therefore, the net effect of rotor fouling is a reduced workload for the motor driver.

Let's now assume that I am driving the same compressor with a gas-fired turbine. The fuel-gas regulator to the turbine is 100 percent open. The turbine is spinning at 10,000 rpm. As the compressor's rotor fouls with salt, what happens to the speed of the turbine?

Answer—it runs faster! It is easier to spin the compressor rotor when its efficiency is impaired. The salt-encrusted wheels do not bite as hard into the gas as would clean wheels. But the amount of gas moved is reduced, even though the rotor is spinning faster.

Jet engine

Did you know that a gas- or diesel-fired turbine driver is essentially the same as a *jet engine?* The burning gas spins a turbine. The turbine spins two compressors:

■ The natural-gas compressor we have been discussing.

■ An air compressor. The discharge pressure from this air compressor might be 80 to 90 psig. The compressed air is used as the combustion air supply to combust the turbine's fuel. The majority of the horsepower output from the turbine (perhaps 60 percent) is used to drive this combustion air compressor.

The horsepower output from a gas turbine is seldom limited by the position of the fuel-gas regulator, as I just described in the previous example. The limit is usually the exhaust temperature of the combustion or flue gases. The turbine's blades have a metallurgical temperature limit of 1100 to 1200°F (as designated by the manufacturer). The temperature of the exhaust combustion gases monitors the temperature of the turbine blades.

Now, let us again assume that our natural-gas compressor rotor begins to foul with salt, drilling mud, and/or a paraffin wax, produced with the gas. Here is what will happen:

■ The flow of compressed natural gas will decrease.

■ The turbine and the compressor will both spin faster.

■ The combustion airflow from the front-end air compressor will increase.

■ The air-to-fuel ratio in the turbine's combustion chamber will increase.

■ The exhaust combustion flue-gas temperature will drop as the air-to-fuel ratio rises.

■ The fuel-gas regulator can now be opened, because we are no longer constrained by the exhaust-gas temperature.

■ The extra fuel gas, plus the extra combustion air, increases the horsepower output from the turbine.

Strange to say, but we could move almost as much natural gas with a dirty compressor rotor as we could with a clean compressor rotor. Of course, the amount of fuel we needed to run the turbine increased substantially as the rotor salted up. But our fuel in Laredo was self-produced, and therefore more or less free natural gas, so we did not care.

Controlling Vibration and Temperature Rise

Vibration

But we learned to care. You see, rotor fouling is a double-edged sword. It cuts two ways. One aspect of rotor fouling is loss of adiabatic compressor efficiency, which wastes work and reduces flow. The other edge of the blade is vibration.

Eventually, the fouling deposits on the rotor will become so thick that they start to break off, especially if you shut the compressor down for a few hours for minor repairs to the lube-oil system. When the compressor is put back on line, bits and pieces of grayish salt break off, and unbalance the rotor. At 8000 rpm, the high-vibration trip cuts off the fuel to the gas turbine, and the machine is taken off line for repair.

The compressor is disassembled. I get the opportunity to accompany the rotor to Dallas, in the back of a van that needed a new suspension. Once there, I watched the manufacturer's machinist crew clean and rebalance the wheels. I noticed that the salt deposits were thickest on the middle wheel. The last wheel was only slightly encrusted while the first wheel was clean, except for some waxy grease.

Why this sort of salt distribution? I reasoned that the entrained brine did not dry out until it reached the middle wheel. But, by the time it reached the last wheel, all the salt deposits that were going to accumulate in the compressor had done so.

This gave me an idea. Suppose we injected a liquid spray into the front end of the compressor (we eventually used a heavy aromatic naphtha, obtained from a local refinery). This could prevent the deposits from sticking to the spinning wheels. We tried it, and it worked. Rotor fouling and the consequent vibrations, and loss of capacity, became far less frequent.

This reminds me of something. Dear reader, you cannot learn anything from our book. You cannot learn about process equipment by reading about it. You have to ride in the back of the van.

Temperature rise

In general, an inefficient compressor will have a high discharge temperature. As compressor efficiency declines, less of the driver's work will go into compression, and more of the driver's work will be degraded into heat.

On the other hand, a high compressor discharge temperature may be due to a larger compression ratio. The compression ratio is the discharge pressure P_2 divided by the suction pressure P_1. Both P_1 and P_2 are expressed in psia (not in psig).

I have put these ideas together in a single equation:

$$\text{Relative efficiency} = \frac{(P2/P1) - 1}{T_2 - T_1} \qquad (30.1)$$

where T_2 is the compressor discharge temperature and T_1 is the compressor suction pressure.

Relative Efficiency

You cannot use the information presented in this text to design compressors. You cannot use the information presented in this book to calculate actual compressor efficiency. Those are complex subjects.

But we are not concerned with establishing the actual compressor efficiency. What we wish to know is the answers to the following sorts of questions:

- What is my compressor efficiency today, compared to its efficiency last month, or right after the unit turnaround last year?

- What is my compressor efficiency today, compared to its design efficiency?

- What is the efficiency of cylinder A, compared to cylinder B, on my reciprocating compressor?

- Which is more efficient—my beat-up, old centrifugal compressor, or my brand-new reciprocating compressor? Both machines are working in parallel, but which has a better adiabatic compression efficiency?

We can answer these questions using Eq. (30.1), which defines *relative efficiency*. The calculated numerical value of relative efficiency means nothing! The equation may be used only to compare two sets of operating data. The equation is not even thermodynamically correct. But it is sufficiently correct, provided the services represented by the two sets of data are reasonably similar.

Axial air compressor example

An axial air compressor is rather like a more simple, but better, version of a centrifugal compressor. Instead of wheels and stators, this model has rotating blades and fixed vanes. In a modern, large jet engine, the combustion air is supplied by an axial air compressor. The machine shown in Fig. 30.1 is supplying air to a fluid catalytic cracker unit catalyst regenerator, in southern Louisiana. The operating parameters shown on the sketch itself were taken when the compressor was thought to be in an abnormally fouled condition. We are going to compare the adiabatic compression efficiency for three conditions:

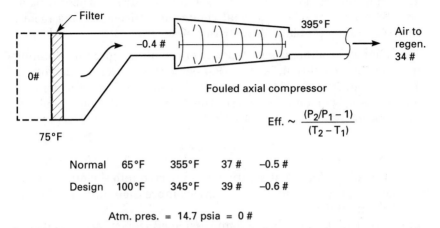

Figure 30.1 Example of relative compression efficiency.

- Fouled
- Normal
- Design

To do this, we will use the relative-efficiency equation [Eq. (30.1)]:

- *Fouled condition:*

$$P_2 = 34 + 14.7 = 48.7 \text{ psia}$$

$$P_1 = -0.4 + 14.7 = 14.3 \text{ psia}$$

$$\frac{P_2}{P_1} - 1 = 2.406$$

$$T_2 - T_1 = 395 - 75 = 320$$

$$\text{Relative efficiency} = \frac{2.406}{320} = .00752$$

- *Normal condition:*

$$\text{Relative efficiency} = .00911$$

- *Design condition:*

$$\text{Relative efficiency} = .01147$$

From these relative-efficiency values, we can draw the following conclusions:

- The axial compressor running in the fouled condition is operating at only 82.5 percent of its "normal condition" efficiency. This means that the fouling problem is really rather severe.

- The adiabatic compression efficiency of the air compressor, running in its normal condition, is only 79.4 percent of its design condition. The manufacturer's design adiabatic compression efficiency is quoted as 81 percent. Therefore, the "normal condition" adiabatic compression efficiency is approximately

$$81\% \times 79.4\% = 64.3\%$$

Not too good! A detergent solution was injected into the air intake of the axial compressor, and the adiabatic compression efficiency recovered to about 80 percent.

I like to calculate the relative compression efficiency because I do not have to know the flow of process gas. I do not have to know the driver horsepower output, the steam to the turbine, the fuel gas to the gas turbine, or the speed of the compressor. I do not have to know Z (the gas compressibility factor) or K (the ratio of the specific heats). The things I do have to know—the suction and discharge temperature and pressure—I can check with my own hands and my own tools.

Parallel compressors

Let us assume that a reciprocating compressor has two cylinders working in parallel. Each cylinder has both a crank-end section and a head-end suction, where gas is compressed. In effect, we have four small compressors working in parallel. The inlet and outlet pressures, and hence the compression ratio, for each of these four minicompressors, is the same. The relative efficiency for each minicompressor is then

$$\text{Relative efficiency} = \frac{1}{T_2 - T_1}$$

You can measure these temperatures with an infrared, noncontact temperature gun. True, this only tells us pipe external temperatures, but as long as we are consistent, use of such skin temperatures is acceptable.

Using this easy technique, we can guide the maintenance effort as to which reciprocating compressor valves must be overhauled, and which valves are working correctly.

Caution: If the inlet side temperature for any cylinder end is hotter than the main gas inlet header pipe, then there is zero gas flow through this cylinder end. Then both the adiabatic and relative compression efficiencies are also zero.

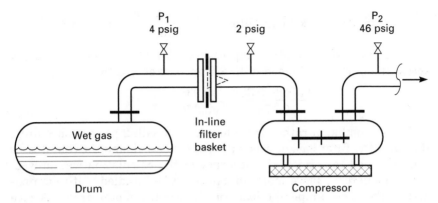

Figure 30.2 In-line basket filter reduces apparent compressor efficiency.

Relative Work: External Pressure Losses

Liz and I recently had a project to expand the wet-gas compressor capacity of a centrifugal machine in Pasadena, Texas. We ran a pressure survey on the compressor system, as summarized in Fig. 30.2.

Our client had been monitoring the inlet pressure to the compressor, at the P_1 pressure point. But we found that the actual compressor inlet pressure was 2 psi lower, due to the in-line basket filter. This filter had apparently never been cleaned. Our client then asked us to calculate the percent of the motor amps being wasted across the partially plugged basket filter.

To answer this question, we may use the following formula:

$$\text{Relative work} \approx \left(\frac{P_2}{P_1}\right)^{(K-1)/K} - 1$$

The ratio of the specific heats K for the gas was 1.33. Therefore

$$\frac{K-1}{K} = \frac{1.33 - 1.00}{1.33} = 0.25$$

Atmospheric pressure in Pasadena, on the day our data were taken, equaled 15 psia, or 30.6 Hg.

Therefore the relative work required by the compressor, with the existing restriction of the fouled filter, was

$$\text{Relative work, fouled filter} = \left(\frac{46 + 15}{2 + 15}\right)^{0.25} - 1 = 0.376$$

The relative work required by the compressor, without the restriction of the fouled filter, would have been

$$\text{Relative work, clean filter} = \left(\frac{46 + 15}{4 + 15}\right)^{0.25} - 1 = 0.338$$

The percent of adiabatic compression work that was being wasted across the partially plugged in-line filter basket was then

$$\frac{0.376 - 0.338}{0.376} = \frac{0.038}{0.376} = 10.1\%$$

This seems like a rather large loss for a small, 2-psi pressure drop. But after the filter basket was pulled and cleaned, and the compressor was returned to service, the accuracy of the calculation was proved. About 10 percent more moles of gas could be handled, with approximately the same amperage load, on the electric motor driver. A nice example of the Second Law of Thermodynamics in action.

31

Safety Equipment

Relief Valves, Corrosion, and Safety Trips

Your process unit likely contains a wide variety of safety features and equipment. These safety features fall under the following three categories:

- Relief valves
- Corrosion monitoring
- Alarms and trips

It is certainly my experience that the most common and catastrophic accidents in process units are related to corrosion-type failures. I cannot actually bring to mind any equipment vessels which were overpressured and failed because a relief valve did not open.

Relief valves were invented to prevent steam boilers from blowing up as a result of excessive steam drum pressure. This was a distressingly common occurrence in the nineteenth century. The relief valve is also called a *safety* (or "pop") *valve*. When the pressure in a vessel exceeds a preset amount, the relief valve is supposed to pop or spring open. Gas will then be vented from the vessel, until the pressure in the vessel drops by 10 to 20 psi below its relief-valve-set pressure. We usually operate pressure vessels 25 psig or 10 percent below the relief-valve setting.

The relief-valve-set, or pop, pressure is adjusted in a machine shop. A large threaded nut on top of the relief valve is used to make this adjustment. Air pressure is applied to the valve, and the technician adjusts this nut until the relief valve opens at the proper pressure.

The pressure that relief valves are set to open is the vessel design pressure. The vessel design pressure, is stamped on the manufacturer's nameplate. It is illegal to set the relief valve at a higher pressure. The vessel is probably hydrostatically tested at a pressure 50 percent greater than its design pressure. This test pressure also is listed on the nameplate. You cannot use the test pressure as a guide to set the relief valves.

Sometimes vessels are rerated. Often, we can increase the design pressure of a vessel by taking credit for an excessive *corrosion allowance,* which is an extra thickness of metal added to the vessel wall. Designers do not include this extra thickness in their pressure-rating calculations. However, if experience teaches that there is little corrosion, then the vessel may be rerated and receive a new pressure-rating stamp. There are rigorous, formal, legal procedures to follow in rerating a vessel.

On the other hand, vessels must often be derated for age and excessive corrosion. Either way, the relief valve needs to be reset. There is only one way to know the actual relief-valve setting, and that is to climb up to the relief valve and read the tag that was fixed to the valve at the time of its last setting.

I used to reset relief valves on small-pressure vessels in the natural-gas fields in southern Texas. After adjusting the nut on the valve with a wrench, I would raise the vessel pressure to check whether the valve would relieve at the proper pressure. If not, I would continue to adjust the nut and, by trial and error, find the proper relief-valve setting. This is not a particularly practical method to adjust relief valves on stream on a process unit. It is also illegal to do so.

Many relief valves pop open above or below their set point. Such valves have to be removed from the vessel and reset in a machine shop. Often, there are isolating block valves, located beneath the relief valve, that permit the relief valve to be pulled while the vessel is still in service. These block valves are perfectly legal, provided they are *chain-locked open.* It in unlawful to have an isolating valve below a relief valve that is not chain-locked open or sealed open in some positive manner.

Relief-Valve Plugging

Often, relief valves do not open because their piping connection to the vessel is plugged with corrosion products, salts, or coke. Even if this connection is only partially plugged, the effective capacity of the relief valve is greatly diminished.

A *rupture disk* is a thin sheet of metal installed below the valve, intended to protect the relief valve from plugging. The rupture disk ruptures at the relief-valve-set pressure. A better approach to retard this plugging problem is to maintain a steam purge, or inert-gas bleed, below the relief valve to prevent the accumulation of solids below the valve.

Often, relief valves fail to close once they have popped open. I hate this! I hate the tension. Will it or won't it reseat by itself? Or, will some operator have to climb up the 180-ft crude distillation column, and hammer on the side of the relief valve until it reseats? Especially if the relief valve is venting to the atmosphere rather than the flare, this can be a nasty and dangerous job.

Furnace tubes, process piping, and heat exchangers may also have to be protected by relief valves. Incidentally, the small $\frac{3}{4}$- or 1-in relief valves you see on many tank field loading lines and on heat exchangers are not process relief valves. They are there for thermal-expansion protection only. This means that if you block in a liquid-filled exchanger and the liquid is heated, the liquid must expand or the exchanger will fail. That is what the thermal relief valve is there to prevent.

Typically, heat exchangers, piping, and furnace tubes have a certain *flange rating*. For example, 150 psi flanges have a pressure rating of about 220 psig, and not 150 psig. A 300 psi flange has a rating of roughly 430 psig. If the upstream process pump has a shut-in or deadhead pressure exceeding this flange rating, then a downstream protective relief valve is required.

Corrosion Monitoring

On older process units, you may still encounter piping with sentry holes. Let's say I have a $\frac{1}{2}$-in thick pipe. The corrosion allowance for the pipe is $\frac{1}{4}$ in. A number of small holes are drilled into the pipe, to a depth of $\frac{1}{4}$ in. When we start leaking at these small holes, this means that the pipe has corroded to its discard thickness in the area of the sentry hole. Incidentally, you can stop the resulting leak, at least in carbon steel water lines, with a brass wood screw and a screwdriver. I have also done this on hydrocarbon lines under an 80 psig pressure, but perhaps that is not too smart.

A more modern method to check for loss of thickness in process piping is by *ultrasonic testing* (UT) or Sonaray. These are portable instruments used to check pipe thickness on-stream. Do not forget, though, that the thin elbow, the one that is sure to fail, is always out of reach unless a ladder can be found. And the inspector cannot find the ladder. The outside radius of elbows are typically the thinnest portions of a piping run. I became the world's leading expert on not finding ladders and not UT-checking thin elbows at the Amoco Oil Refinery in Texas City, in 1976—just prior to the alkylation unit explosion.

Corrosion coupons

I always like corrosion coupons. They are a tangible piece of hardware that you can hold in your hand. A corrosion coupon is just a piece of metal, perhaps $\frac{1}{2}$ in thick, 3 in long, and $\frac{1}{2}$ in wide, that is inserted into the flowing process stream through a packing gland in a valve. It can be easily pulled for inspection every week or every month. The corrosion engineer weighs the coupon for metal loss and inspects it for pitting and other forms of corrosive attack. Often, a number of corrosion coupons of various metallurgy are used. The corrosion engineer then calculates, on the basis of the metal lost from the coupon in one month, the *mils per year* corrosion rate.

For example, a steel pipe is $\frac{1}{2}$ in thick, or 250 mils. Its discard thickness is $\frac{1}{8}$ in thick, or 125 mils. If the corrosion rate is measured at 25 mils per year, then the expected life of the pipe is 5 years. Corrosion rates in excess of 10 mils per year are normally considered excessive and unacceptable, at least in petroleum refineries.

Corrosion probes

This is an electronic method used to measure corrosion rate in mils per year. The corrosion probe can be inserted through a packing gland. It is read periodically with a portable instrument that measures the change in electrical conductivity of the probe. It is simple, but perhaps a little less reliable than the coupon.

There is another sort of probe that measures ionic hydrogen penetration into and through steel walls. Such *hydrogen activity* is a product of corrosion and/or high hydrogen partial pressures in a vessel. This sort of hydrogen activity promotes hydrogen blistering and stress corrosion cracking of vessel walls. A sketch of a hydrogen probe is shown in Fig. 31.1. It consists simply of a small chamber welded onto the exterior of a vessel. A pressure gauge is connected to this chamber.

The hydrogen ions (or protons) dissolve, and pass through the iron lattice structure of the vessel wall. When some of them emerge at the outside of the vessel wall, they are trapped inside the hydrogen probe chamber. There, the ionic hydrogen is converted into molecular hydrogen. The rate of pressure increase inside the chamber is a direct measure of the amount of ionic hydrogen infiltration through the lattice structure of the steel wall of the process vessel.

Often massive corrosion failures occur suddenly, without warning of any small leaks. Lines part at welds, vessels burst apart as a result of

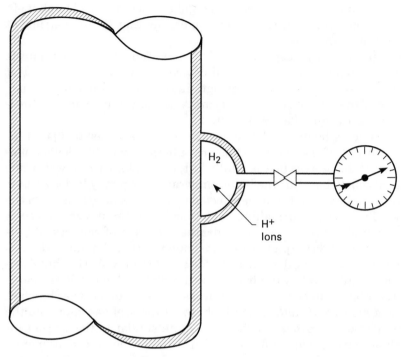

Figure 31.1 A hydrogen probe.

hydrogen-assisted stress corrosion cracking, and thin elbows peel back like the top of a soup can. Process plants are dangerous places, mainly because of corrosion; hence the importance of monitoring corrosion. It is the responsibility of the unit chemical or process engineers to monitor corrosion on their units.

Alarms and Trips

Safety trips

One of the most common safety trips is the automatic fuel-gas shutoff. We have this at home on our furnaces. We ignite the pilot light manually. The pilot light heats a thermocouple. The milliamp output from this thermocouple opens the fuel-gas valve to the main burner. The gas in the main burner is ignited from the pilot-light flame. Should the heat from the pilot light diminish below a certain point, the fuel gas to the main burner, as well as to the pilot light, will be shut off.

A less common type of fuel-gas trip to a heater is a low-pressure trip. A pressure transducer generates a milliamp output from a boiler feedwater pump. Should this milliamp output fall below a certain level, the instrument air signal to the fuel-gas regulator actuator will be shut off. These fuel-gas valves are *air to open,* meaning that loss of instrument airflow causes the valve to close.

Some fired heaters, especially boilers, have a device called a "purple peeper," which is simply an optical device that looks at a flame. If it does not detect light with a wavelength in the high-frequency (i.e., purple) end of the optical scale, it interprets this as a flame-out. The fuel-gas regulator is automatically shut.

High process heater outlet temperatures are another trip point for many heaters. These sorts of trips are subject to a rather deadly malfunction. If the process flow is totally lost, the heater tubes may become extraordinarily hot. But the high-temperature trip may be located too far from the heater outlet to be affected by the high temperature in the heater. This particular malfunction occurred in a southern Louisiana refinery, and indirectly led to the deaths of a score of workers. This is why we have P&ID (piping and instrumentation diagram) reviews when our unit is being designed. Fuel to a boiler, may also be shut down by a low water level in the boiler's steam drum. The low-level trip is simply a float chamber. If the float drops to the bottom of the chamber, it flips a *mercuroid switch,* which shuts off the fuel-gas supply to the boiler. These mercuroid switches look quite similar to the thermostat switch we have inside our house, which switches the air conditioner on or off. Note that these level trip float chambers, even though they are quite short, still require two level taps on the vessel.

Some heaters also have a low fuel-gas pressure trip on the fuel gas itself. The idea here is that if fuel-gas flow is lost, we do not want it to surge back into the heater too quickly, if it is suddenly restored.

Compressor trips

Our home circuit breakers, or fuses, are, of course, trips to prevent overheating electric circuits or electric motors. The only difference is that at work, our electric circuit breakers have a built-in time delay. This is needed to allow the motor driver to overcome the starting torque inherent in most large pieces of rotating process equipment.

Compressors also have vibration trips. These trips measure the amplitude of the vibrations—which, if they become excessive, will shut off the fuel, steam, or electricity to the compressor's driver.

Turbines, both gas and steam, also have overspeed trips. These consist of a little flywheel, constructed from three balls. The little balls are spread apart by centrifugal force. The greater the rpm, the greater the centrifugal force. If the balls spread apart too far, they activate the trip. We have James Watt to thank for this neat innovation, still used in its original form.

Most large compressors also have a low-lube-oil-pressure trip. This again would shut off the fuel or steam to a turbine, if the lube-oil pressure gets too low. It is interesting to note that the low-lube-oil pressure that activates the trip is not the lube-oil pressure to the bearing that would be damaged because of a lack of lube-oil flow. Rather, it is the low-lube-oil pressure to the trip switch itself that directly shuts down the turbine or motor. Thus, with a low-lube-oil-pressure, one can trip off a compressor without actually losing lubrication flow to the bearings at all.

Compressors are also served by high-liquid-level trips in their upstream knockout drums. These high-liquid-level trips work in the same way as the low-level boiler trips discussed above, except that the mercuroid switch is activated by a rising, rather than a falling, liquid level so as to protect the compressor from a slug of liquid.

A final word about trips. Any trip that is not tested on some routine basis can never work in an emergency. I will guarantee you that when the coupling shears on that steam turbine, driving that 4000-bhp compressor, that overspeed trip that should shut off the steam flow to the turbine will not in fact trip if you have not tested that same trip recently. I promise you that the trip valve mechanism will be encrusted with hardness deposits from the steam. While the trip lever may be unlatched, the turbine will continue to spin merrily along, until it self-destructs as a result of uncontrolled overspeed. And, ladies and gentlemen, you may imagine how I have become so knowledgeable about this particular subject.

If you have any questions about the information presented in this book, please feel free to call or fax us in New Orleans.

- Fax (504) 456-1835
- Phone (504) 887-7714

Thank you so very much for reading our book.

Liz and Norm Lieberman

Glossary

actual level   This is the liquid level actually in the vessel.

adiabatic combustion   Burning fuel with no radiant heat losses.

adiabatic process   No heat lost or gained.

aeration factor   The actual density of a froth, divided by its clear liquid density.

affinity or fan laws   Relates speed of centrifugal machines to head, flow, and work.

afterburn   Reignition of flue gas in the convective section, of a fired heater.

air to open   Control valves are operated by air pressure. A control valve that is opened by air pressure is "air to open."

apparent level   This is the liquid level that appears in the gauge glass, or is displayed on the control panel.

available NPSH   The difference between the pressure at the suction of a pump and the vapor pressure of the liquid.

barometric condenser   Condensation under vacuum by direct contact with the cooling water.

blown condensate seal   Vapor or steam, flowing through the condensate outlet nozzle, of a condenser.

boiler codes   Laws describing the safe design for boilers and other pressure vessels.

boiler feedwater   Water that has been softened or demineralized, and deaerated.

bottom dead center   The piston position in a cylinder, when it is closest to the crank shaft.

carbonic acid   Corrosive CO_2 dissolved in water.

cavitation   Flashing of liquid in a confined space, such as a nozzle.

centipoise   The most common measure of viscosity.

centistokes   A measure of viscosity.

chain-locked open   The practice of passing a chain through a valve handle and locking it in place, to prevent it from being closed without the proper authority.

channel head   The portion of a shell-and-tube heat exchanger that covers the channel head tubesheet.

channel head cover   A round plate that bolts onto the channel head.

channel head pass partition   The baffles which segment a heat-exchanger channel head, to create multipass exchangers.

channel head tubesheet   The fixed end of the tube bundle.

channeling   Loss of contacting efficiency, due to uneven flow distribution.

clear liquid   Density of liquid assuming no foam.

cogeneration   Use of turbine exhaust steam, to provide process heat.

compression ratio   The discharge pressure divided by the suction pressure.

compression work   Energy provided to a gas, to increase its pressure.

condensate drum balance line   Used to connect the channel head of a steam reboiler to the condensate drum, for pressure balance.

condensing turbine   Exhausts into a surface or barometric condenser running at a vacuum.

convective heat   The sensible-heat content change of a flowing fluid.

corrosion allowance   Amount of extra metal thickness added to a pipe or vessel wall, over and above that required for pressure to allow for future corrosion.

crank end   The end of the cylinder closest to the crank shaft.

critical flow   Vapor flowing at the speed of sound.

critical speed   The natural harmonic frequency of a centrifugal compressor, which determines its critical, or self-destructive speed.

cross-flow velocity   Fluid flow at right angles to a heat-transfer tube.

deaerator   Used to remove oxygen from boiler feedwater.

degree of fractionation   A measure of how much of the light components are left in the bottoms product, and how much of the heavier components are left in the overhead product.

demister pad   A steel, plastic, or straw mesh pad used to coalesce droplets of liquid.

dependent variables   Parameters which an operator does not have direct control of.

dew-point equation   Used to predict when a vapor will start to condense.

downcomer seal   On a distillation tray, the height of liquid above the bottom edge of the downcomer.

draw-off boot   A low point in a vessel, used to collect water.

drip points   On a packed bed, this refers to the number of points that liquid enters the top of the bed.

driver-limited   A pump which is limited by the horsepower of its turbine or motor.

drying out   Liquid flowing down trays in a tower progressively evaporating.

dynamic machines   Pumps or compressors that convert velocity into feet of head.

elevation head loss   When liquid flows uphill, it loses pressure due to the increased elevation.

emulsion   A mixture of two, insoluble phases; like oil and water.

entrainment coefficient   A measure of the tendency of a flowing gas to carry droplets of liquid.

entropy   A measure of a fluid's ability to do work.

equilibrium conditions   Vapor and liquid in intimate contact are said to be in equilibrium.

excess O2   A measure of the amount of excess air used in combustion.

expansion joint   A flexible section of piping with bellows, which permits thermal expansion.

explosive region   The ratio of fuel to oxygen that will sustain combustion.

extended surface   Fins, studs, or serrations on the outside surface of heat-exchanger tubes.

eye of the impeller   The inlet nozzle in a centrifugal pump's impeller.

fan blade pitch   The angle of a fan blade. The bigger the pitch, the greater the airflow.

fan laws   Relationship between speed, head, and flow for centrifugal machines.

feed preheater   The heat exchanger used to increase the enthalpy of the feed to a distillation tower.

film resistance   The property of a fluid that diminishes heat exchange.

fin tube bundle   Serrated, or extended-surface, exchanger tubes.

FLA   Full-limit amps, or the high-amperage trip point for a motor.

flange rating   Piping and heat-exchanger flanges have a nominal pressure rating less than their actual pressure rating.

flash points   Temperature at which a liquid will ignite when exposed to a flame.

floating head   The component of a shell-and-tube exchanger that covers the floating-head tubesheet.

floating-head exchanger   Shell-and-tube exchanger in which one end of the tube bundle is left free to move, to accommodate thermal expansion.

floating-head tubesheet   The end of a tube bundle left free to move, due to thermal expansion.

flooded condenser control   A method of pressure control in a distillation tower.

flooding   In distillation, the point in the operation where increases in vapor-liquid loading cause a noticeable reduction in the fractionation efficiency or tray efficiency of the tower.

forced circulation   Circulation in reboilers which receive their liquid feed from the discharge of a pump.

frictional loss   Pressure reduction accompanied by heat generated by a fluid flowing through piping.

gauge glass   Used to measure the pressure difference between two points in a vessel, in terms of the density of liquid in the glass.

governor speed-control valve   Controls a turbine's speed automatically by adjusting the steam flow into the steam chest.

head end   The end of the cylinder farthest away from the crank shaft.

header box   A box from which a number of pipes emerge.

head loss   Pressure lost due to friction, in a flowing liquid.

head pressure   The pressure exerted by a height of liquid.

heat balance   Heat inputs have to equal heat outputs and ambient heat losses.

heat flux   Heat input, per unit of time, per unit of heat-transfer surface area.

heat-proving the heater   Optimizing airflow, to maximize combustion efficiency.

heat-transfer surface area   The outside surface area of a tube used to transfer heat.

HETP   Height equivalent to a theoretical plate (or tray). Also called an *equilibrium separation stage*.

high-temperature creep   The plastic deformation of steel, at high pressure and high temperature.

holdup   The time that liquid is retained in a vessel; usually to promote separation or provide hold time for a downstream pump.

hogging jet   A single stage jet, exhausting to the atmosphere.

hot well   The drum located below a surface condenser in vacuum service, where the barometric legs drain to.

hot spot   The glowing area on a heater tube.

hydraulic hammer   The correct term for water hammer.

hydraulics   The study of fluid flow, pressure, and feet of head.

hydrogen activity   The tendency of ionic hydrogen, or protons, to dissolve into the lattice structure of steel walls.

ideal compression work   Compression work discounting friction, leakage, and other mechanical losses.

ideal-gas law   An equation describing the theoretical relationship between the pressure, temperature, and volume of gas.

impingement plate   Used to protect tubes from the erosive velocity of the shell-side inlet fluid.

incipient flood point   The vapor-liquid loading, at which a distillation tray develops its maximum efficiency.

independent variables   Parameters which an operator has complete control to change.

indicator card   A record of the pressure vs   volume of gas inside the cylinder of a reciprocating compressor.

intake valves   The inlet port valves of a reciprocating compressor.

internal reflux   The liquid flow leaving a distillation tower tray.

irreversible process   A machine that converts the pressure of a fluid into work always converts some of the potential work into heat. This heat is a measure of the irreversibility of the process.

isoenthalpic   An expansion of steam which does not involve any energy loss from the steam.

isoentropic   A process change, where the ability of the system to do work has not been reduced. A reversible process.

jet performance curve   Issued by the steam jet manufacturer, describing the vacuum a jet should pull, at various gas flow rates.

kettle reboiler   A type of reboiler which depends on the force of gravity for its liquid feed.

laminar flow   Fluid flow with little turbulence.

latent heat   Heat needed to change a liquid at its boiling point, into a vapor at its dew point.

latent-heat transfer   Heat exchange involving vaporization or condensation only.

letdown valve   Used to reduce the pressure of a liquid refrigerant.

lift gas   Vapors used to lift solids or liquids to a higher elevation.

LMTD   The natural log-mean temperature difference that promotes heat transfer.

loop seal   Used to prevent vapor from blowing back through a pipe, to an area of lower pressure.

mass flow   Pounds of flow.

metering pump   Used to both pump and measure the flow rate of small chemical additive streams.

milliamp output   The small (milliampere-range) electrical output from a thermocouple or pressure transducer that is used for control purposes.

mils per year   One mil equals 0.001 in. Mils per year is a measure of how fast a steel wall corrodes away.

molecular weight   The weight of a gas, divided by the number of moles of the gas.

Mollier diagram   A chart relating enthalpy, entropy, temperature, and pressure of steam.

motor pulley   The wheel fixed to the shaft on a motor, used to turn a belt drive.

nozzle exit loss   Conversion of pressure into velocity.

NPSH   Net positive suction head.

open area   In a packed bed, the open area available for vapor flow.

out-of-level   The difference between the low point and high point of a distillation tower tray.

overamping   Tripping a motor because of excessive electrical power demand. Exceeding the FLA limit.

partial pressure   The vapor pressure of a component, multiplied by its mole percent concentration.

P&ID   Piping and instrumentation diagram(s): the drawing(s) that most completely describe the process construction of the plant.

phase rule   Fundamental concept relating the number of independent variables in a process.

polar molecule   A molecule, like water, that can conduct electricity.

pressure recovery   Conversion of velocity into pressure.

pressure transducers   A device that converts pressure into a small electrical flow.

prime   Filling the case of a pump with liquid.

pulsation   An acoustical phenomenon which causes a variable pressure inside the cylinder of a reciprocating compressor.

pumparound   A method to remove heat from a section of a tower.

pumparound draw tray   The tray from which liquid is drawn off to the pumparound pump.

pumparound return tray   The tray below the pumparound return nozzle.

ratio of the phases   A concept in level measurement, in which the relative heights of gas, hydrocarbons, and water must be considered.

reboiler   Source of heat input to a distillation tower.

redistributor   A plate that takes vapor or liquid flowing through a tower and, by means of orifices, forces the fluid to flow more evenly through a packed bed.

reflux comes from the reboiler   Concept that the reflux flow originates from heat supplied from the reboiler.

relative efficiency   The efficiency of a compressor, as compared to the efficiency of the same machine at a different time.

relative volatility   The ratio of the vapor pressure of the light component, divided by the vapor pressure of the heavier component, at a particular temperature.

required NPSH   The net positive suction head needed to keep a pump from cavitating.

reversible process   Work extracted from the velocity and pressure of a fluid, which can be put back into the fluid to restore its pressure and temperature. An isoentropic process.

rotating assembly   The portion of a centrifugal machine that spins.

rotor   The spinning part of a centrifugal compressor or pump.

running NPSH   The required liquid head needed to convert into velocity, as the liquid flows into a centrifugal pump.

saturated liquid   Liquid at its boiling point.

saturated steam   Steam at its dew point.

saturated vapor   Vapor at its dew point.

secondary ignition   Reignition of the flue gas in the convective section of a heater.

self-flushed pumps   A centrifugal pump that obtains its seal flush liquid, from the pump's own discharge.

sensible heat   Heat content due to temperature.

sensible-heat transfer   Heat exchange without change of phase.

stab-in reboiler   A reboiler tube bundle inserted directly into the base of a tower.

starting NPSH   The required liquid head needed to overcome the inertia of static liquid in the suction of a centrifugal pump.

starting volumetric clearance   The space between the piston and cylinder header, at the end of the stroke, in a reciprocating compressor.

stator   The stationary component on a centrifugal compressor.

steam chest   The portion of a steam turbine between the governor inlet valve and the turbine case.

steam rack   Device that automatically opens and closes steam nozzles into a turbine case.

steam tables   A table summarizing the enthalpy, density, pressure, and temperature of steam.

steam trap   Used to drain water from a steam system.

stuttering feed interruption   Erratic flow of liquid into a heater.

subcooled   Liquid below its boiling point.

sublimation   The change of a solid directly to a vapor—or the reverse.

surface condenser   A condenser having separate vapor and liquid outlets.

surge   An unstable and dangerous operating mode of a centrifugal compressor.

temperature difference   In distillation, the tower bottom temperature minus the tower top temperature.

the point of absolute combustion   Air rate required to maximize heat recovery into the process, and minimize heat lost up the stack.

thermowell   The casing used to protect the thermocouple wires.

top dead center   The position of a piston in a cylinder, when it is closest to the cylinder head.

top-tray flooding   Tower flooding caused only by flooding on the top tray.

total dissolved solids (TDS)   A measure of the tendency of boiler feedwater to form scale on a boiler's tubes.

tramp air   Air accidentally introduced into a firebox, but not through the burners.

tray spacing   The vertical spacing (typically 24 inches) between tray decks in a distillation tower.

tube pitch   Space between the centers of adjacent tubes in a heat exchanger.

tube skin temperature   The exterior temperature of a heater tube.

tube support baffles   Used to support tubes in a shell-and-tube heat exchanger.

turbine case   The portion of the turbine containing the wheel.

turndown   The minimum percentage of the design capacity at which process equipment will operate, and still maintain a reasonable degree of efficiency.

valve disablers   Used in reciprocating compressors to reduce capacity.

valve ports   Openings in a reciprocating compressor's cylinder that permit the entry, or exit, of the flowing gas.

velocity boost   Pressure increase in a steam jet, due to the reduction of velocity.

velocity steam   Used to increase the velocity of a liquid in a heater tube.

vertical baffle   Used in the bottom of distillation towers to segregate the cold-and-hot zone, for more efficient reboiling.

volumetric efficiency   A measure of the volume of gas moved per stroke in a reciprocating compressor.

vortex breaker   A mechanical device, set over an outlet nozzle, to prevent drawing of vapor out of the nozzle, with the flowing liquid, due to the formation of a vortex.

waste-heat boiler   Used to generate steam from excess process heat.

water hammer   The sudden velocity reduction of a flowing liquid is converted into pressure surges which then produce water hammer.

water rate   Pounds of steam needed to generate one horsepower (hp) in a steam turbine.

weir loading   On a distillation tray, the GPM (gallons per minute) of liquid flow, divided by the inches of outlet weir length.

wetted surface area   In a packed bed, the area of the packing exposed to the internal reflux. A measure of the packing's vapor-liquid contacting efficiency.

The Norm Lieberman Video Library of Troubleshooting Process Operations

This video library provides a clear explanation of the function of process equipment, and explains how to optimize operations and identify malfunctions. It is intended for process engineers as a source of reference and for experienced operating personnel as advanced training. Personnel at all levels can use this library to improve process operations. Tapes may be purchased singly.

This program is an extension of *A Working Guide to Process Equipment* and other books available through the authors. For further information, contact the authors directly at

Process Chemicals, Inc.
 5000A West Esplanade, Suite 267
 Metairie, LA 70006 (USA)
 Phone (504) 887-7714
 Fax (504) 456-1835

Index

Absolute combustion, 252–256
Absolute manometer, 192–193
Absolute pressure, 28
Absorption section, 39–40
Acceleration, pump suction, 330
Adiabatic combustion, 277
Adiabatic compressor efficiency, 383
Adiabatic flame temperature, 281
Adiabatic process defined, 356
Aeration factor, 9, 18
Aerated liquid, 9, 17–18
Aerodynamic stall, 365–366
Aerodynamics, lift in, 365–366
Affinity Law, 167
Afterburn, 252–254, 257, 271
Air compressor, 375
Air coolers, 104, 163
Air leaks, 105
    in heaters, 260, 271
    surface condenser, 226
Air-fuel mixing, 266–267, 288
Air lock, 149
Air preheater, 268, 272
Air register, 251, 263
Airlift pump, 45
Alarms, 401–403
Ambient heat losses, 117, 275
American Petroleum Institute, 114
Amine:
    regeneration, 237
    salts, 387
Ammonium:
    chloride, 171, 387
    sulfide, 200–201
Amp load on motor, 320
Amperage meter, 166

Annular flow, 287
Aromatic steam stripping, 122
Ash deposits, heater tubes, 284
Axial compressor, 392

Backflowing air cooler, 166
Backmixing, 51
Baffle cut, 233–234
Balance line, 95
Barometric condenser, 215–217
Barometric pressure compensation, 65
Beam engine, 217
Bearing Buddy, 321
Bearings, 307–308, 321
Belt drives, 167
Benzene stripping, 119
Bernoulli's Equation, 126, 302–304
Beta scan plot, 384
Blown condensate seal, 89
Boiler, 345
    blowdown from, 180
    feedwater, 99–101, 174–176
Bottom dead center, 379
British Thermal Unit, BTU, 357
Bubble cap, 17
    riser, 21
    trays, 7, 21–22
Bubble point, 107–109, 130, 306,
    328–330
Burner, 266–268

Calcium carbonate deposits, 138
Carbon dioxide corrosion, 175
Carbonates, 149, 174

417